Sustainable and Collaborative Tourism in a Digital World

Advances in Tourism Marketing

Metin Kozak, Antónia Correia and Alan Fyall

The purpose of this series of cutting-edge research-informed edited books is to introduce the reader to a range of contemporary marketing phenomena in the domain of travel and tourism. Authored by leading academics in their fields of research interest, each book will bring together a selection of related themes with individual chapters contributing theoretical, methodological, policy-related and/or practical outcomes for the reader. Each book will be introduced and brought to a conclusion by the series editors who between them have many decades of research and publishing experience. The singular aim of this advanced series of scholarly texts is to stimulate and engage readers in the fast-changing, complex and increasingly interdisciplinary nature of tourism marketing, and serve as a catalyst for future intellectual, academic, and professional-driven research agendas. This series encourages critical, participatory and humanistic approaches to research and welcomes contributions from all over the world. In particular, the series welcomes contributions from a non-Western perspective as tourism becomes truly global in both its reach and impact.

Sustainable and Collaborative Tourism in a Digital World

Editors:

Alain Decrop, Antónia Correia, Alan Fyall and Metin Kozak

(G) Goodfellow Publishers Ltd

(G) Published by Goodfellow Publishers Limited,
26 Home Close, Wolvercote, Oxford OX2 8PS
http://www.goodfellowpublishers.com

British Library Cataloguing in Publication Data: a catalogue record for this title is available from the British Library.

Library of Congress Catalog Card Number: on file.

ISBN: 978-1-911635-76-5

DOI: 10.23912/9781911635765-4477

Design and typesetting by P.K. McBride, www.macbride.org.uk

Cover design by Cylinder

Contents

Part 4: Technology, residents and over-tourism

List of tables

List of figures

Author biographies

Alain Decrop, is Professor of Marketing at the University of Namur and President of the French Marketing Association (AFM). He is the Director of CeRCLe (Center for Research on Consumption and Leisure) and the co-Director of CCMS (Center on Consumers and Marketing Strategy). His current research interests focus on the sharing economy, decision-making processes and contemporary consumption phenomena, with applications most often related to tourism and leisure. He has published a large number of books, chapters and articles in top-tier journals around these topics. E-mail: alain.decrop@unamur.be

Antónia Correia, Ph.D., is Professor of Tourist Behaviour and Tourism Economics, University of Algarve, and at the Tourism and Hospitality School at Universidade Europeia, in Portugal. Her research areas include consumer behaviour, tourism economics and modelling. She has published more than 100 papers in tourism, leisure and economics journals and is on the editorial boards of several, including *Journal of Travel Research, Journal of Business Research, Tourism Analysis* and *Anatolia*. E-mail: ahcorreia@gmail.com.

Alan Fyall, Ph.D., is Associate Dean Academic Affairs and Visit Orlando Endowed Chair of Tourism Marketing at the Rosen College of Hospitality Management, University of Central Florida. Alan has published widely in the areas of tourism and destination marketing and management including 22 books and 150 articles and book chapters. He is Editor of Elsevier's *Journal of Destination Marketing & Management* and sits on the editorial boards of many leading journals. E-mail: alan.fyall@ucf.edu.

Metin Kozak, Ph.D., is Professor in the School of Tourism, Dokuz Eylul University, Turkey. His research focuses on consumer behaviour, destination management, and sustainability. He is the co-editor of *Anatolia: An International Journal of Tourism and Hospitality Research* and has been to several universities in the U.S., Europe and Asia as a visiting scholar and many conferences as an invited speaker. E–mail: M.Kozak@superonline.com.

José Luís Abrantes, Ph.D. is Professor of Management and Marketing, Head of Department at the School of Technology and Management, and the Coordinator of CISeD (Centre for Research in Digital Services) of the Polytechnic Institute of Viseu. His research areas are marketing, tourism management and pedagogy. He has published more than 100 papers in

Annals of Tourism Research, Tourism Management, Journal of Business Research, European Journal of Marketing, International Marketing Review, amongst others. E-mail: jlabrantes@estgv.ipv.pt.

Arash Akhshik, Ph.D., is a lecturer at the Faculty of Tourism, Eastern Mediterranean University, Famagusta, Turkey. He has published various articles in peer-reviewed top-tier journals and has presented his research at several international conferences on tourism and hospitality management. His research interest includes visitor management and experience design. E-mail: arash.akhshik@emu.edu.tr.

Marcello Atzeni, PhD, is a research fellow at the Department of History and Human Science at the University of Sassari (Italy). His research interests include tourism marketing, cultural and heritage tourism and authenticity in tourism. E-mail: matzeni1@uniss.it.

Stéphane Bourliataux-Lajoinie, Ph.D., is senior lecturer at Conservatoire National des Arts et Métiers, Paris (France), and a member of LIRSA. He has published many research papers and book chapters on digital, online consumer behaviour and privacy. He is pedagogical head of Masters E-business and digital marketing in Paris. He has been a visiting professor in different countries including Brazil, Lebanon, India, and Africa. E-mail: stephane.bourliataux@lecnam.net.

Silvana Canales Gutiérrez, Ph.D. studied at Rovira i Virgili University in Tarragona, Spain, with a master's degree in business and contracting law from the same university, and master's degree in Law with an investigative focus from the National University of Colombia. Her research areas include private international law, consumer rights, collaborative economy especially in the tourism sector, abusive clauses and cryptocurrencies. Email: silvana.canales@urv.cat

Francesca Checchinato, Ph.D., is Associate Professor of Marketing at the Ca' Foscari University, Venice. Her research is focused on brand management, digital marketing and communication. She is a co-founder of the Agrifood Management & Innovation Lab and scientific coordinator of its digital marketing research area. She teaches across a range of postgraduate and executive programs in marketing and branding strategy. E-mail: f.checchinato@unive.it.

Giacomo Del Chiappa, Ph.D, is Associate Professor of Marketing at the University of Sassari (Italy) and Senior Research Fellow at the School of Tourism & Hospitality of University of Johannesburg (South Africa). His research areas include destination governance and branding, sustainable tourism, consumer/tourist behaviour, digital marketing and airport marketing. He is author and co-author of around 200 national and international publications and member of the editorial boards of several leading journals. E-mail: gdelchiappa@uniss.it.

Josep Lluis del Olmo Arriaga, Ph.D., is Professor of Marketing at University of Abat Oliba CEU in Barcelone (Spain). Her research areas include tourism marketing, market research and social media. She has published more than 50 papers in marketing, tourism and social media journals, and chapters on tourism marketing. She is a reviewer for several marketing journals, such as *Journal of Global Fashion Marketing*. E-mail: jlolmo@uao.es.

Marie Dewitte, Ph.D student, is researcher in marketing at the University of Namur. Her main research interests are consumer behaviour, sustainable marketing and green consumption. She is a member of the Namur Digital Institude (NADI) and the Center for Research on Consumption and Leisure (CeRCLe). E-mail: marie.dewitte@unamur.be.

Nico Didry, Ph.D, is Professor of Marketing and Consumer Behavior at University Grenoble-Alpes, where he leads the Sport & Tourism Economical Strategies program. His research focuses on leisure experiential consumption, tourism and territorial strategies with an ethnomarketing approach. He is a member of expert committees of tourism organizations and of the economics research center of Grenoble University. e-mail: nico.didry@univ-grenoble-alpes.fr.

Frédéric Dosquet, Ph.D. is Professor at ESCPAU Business School. He has published more than 40 contributions mainly on tourism, public management and political marketing. He is visiting professor at University Saint Joseph of Beirut (Lebanon) and Lagunes University (Ivory Coast), and a columnist in French newspapers, radios and TV. E-mail : frederic.dosquet@esc-pau.fr.

Jean-Luc Giannelloni, Ph.D., HDR, is full Professor of Marketing and Consumer Behavior at University Grenoble-Alpes. His research interests mainly lie in responsible consumption, hospitality and tourism. He has published in the *International Journal of Research in Tourism* and *European Journal of Tourism Research*, amongst others. He is currently Vice-President for publications of the French Marketing Association and former co-editor in chief of *Décisions Marketing*, a peer-reviewed academic journal published by afm. e-mail: jean-luc.giannelloni@univ-grenoble-alpes.fr.

Xing Han received her Ph.D. degree in Modern Commerce from Otaru University of Commerce. Her research interests include consumer behavior, tourist social interaction, and tourism co-creation. E-mail: xinghanida@hotmail.com.

Foad Irani, Ph.D. candidate of Tourism Management, is a research assistant at the Faculty of Tourism at Eastern Mediterranean University, Famagusta, TRNC, Turkey. He has presented his research at several international conferences on tourism and hospitality management. Research areas include consumer behaviour, sustainable development, and human resource management. E-mail: foad.irani@emu.edu.tr.

Stella Kladou, PhD., is a Postdoctoral researcher at the Hellenic Mediterranean University. She has served as an Assistant Professor in the Tourism Administration Department at Boğaziçi University in Istanbul, Turkey and a Senior Lecturer at Sheffield Hallam University in the UK. She has published in prominent academic journals and books of international publishers, and presented at well-esteemed international conferences. Her research interests include place branding, consumer behavior, cultural tourism, and digital marketing. E-mail: stellakladou@hmu.gr

Sina Kuzuoglu is a PhD student at the Department of Recreation of Leisure Studies at the University of Waterloo in Canada. He has worked on a number of World Heritage Sites in Turkey and his M.A. thesis focused on the relationship between tourism and conservation with the subjective well-being of residents in Istanbul's historic city center. His research interests include heritage cities, tourism planning and governance, urban communities, and health and well-being. Email: skuzuogl@uwaterloo.ca

Jérôme Mallargé, Ph.D., is a researcher in marketing at the University of Namur. His research interest is mainly on collaborative consumption, consumer behaviour and consumer evaluation process. He is a member of the Namur Digital Insitute (NADI) and of the Center for Research on Consumption and Leisure (CeRCLe). E-mail: jerome.mallargé@unamur.be.

Michelle Moraes, Post-doctorate in Economics from the University of Aveiro (Portugal) and in Political Science and International Relations from the University Fernando Pessoa (Portugal), holds a PhD in Economics from the Università degli Studi di Ferrara (Italy) and in Public Policy from the Federal University of Maranhão (Brazil). She has published several articles on innovation, tourism and sustainability, and is on the board of several large companies in the service sector in Brazil and Italy. E-mail: michelle.moraes@universidadeeuropeia.pt.

Ali Öztüren, Ph.D., is a Professor of Tourism Management in the Faculty of Tourism at the Eastern Mediterranean University and currently Vice Dean of the Faculty. He has published various articles in peer-reviewed scholarly journals and has presented his research at several international conferences on tourism and hospitality management. His current research interests include tourism and hospitality management, sustainable tourism management, service operations, and quality management. He is on the executive boards of various national and international academic projects and events, on the referee boards of many international and national journals, and on the advisory boards of academic and social organizations. E-mail: ali.ozturen@emu.edu.tr.

Andreia Pereira holds a master's degree in Tourism Management from the Polytechnic Institute of Viseu and is a PhD student in Tourism, Heritage and Territory at the Faculty of Letters of the University of Coimbra. Her main research interests are the study of storytelling, generations, gender in tourism. E-mail: arapereira@outlook.pt.

Marina A. Petruzzi, is Ph.D. Candidate in Tourism Management at the Instituto Universitário de Lisboa, Business Research Unit, Lisbon, Portugal and Unidade de Investigação em Turismo e Hospitalidade, Faculdade de Turismo e Hospitalidade, Universidade Europeia, Lisboa, Portugal. Her research areas include peer-to-peer accommodation in the sharing economy and tourism sustainability. E-mail: marina.a.petruzzi@gmail.com.

Carolus Praet is Professor of International Marketing at Otaru University of Commerce, and Guest Professor (Part time) at Keio University's Business School and at Kobe University's Graduate School of Management SESAMI program, Japan. His research interests include cross-cultural advertising and consumer behavior, services marketing, and tourism marketing. He has authored over 30 articles and book chapters on advertising and marketing in English and Japanese. He is a member of the editorial boards of *International Journal of Marketing & Distribution* and *Journal of Advertising Science.* E-mail: praetcar@res.otaru-uc.ac.jp.

Manuel Reis, PhD candidate in Management in the University of Beira Interior, is Assistant Teacher at the Polytechnic Institute of Viseu – Higher School of Technology and Management. He has a BSc is in Mathematics from the New University of Lisbon and an MSc in Statistics from the University of Porto. He is affiliated with the Center for Studies in Education, Technologies and Health. E-mail: manuelreis@estgv.ipv.pt.

Áurea Rodrigues has a PhD in Tourism from the University of Aveiro. She is an assistant professor in Tourism Studies at the University of Évora and an affiliate researcher at CIDEHUS, UÉ, Portugal. Her main research interests focus on sustainable rural tourism development, ecotourism, astro-tourism, consumer behavior in tourism and tourism marketing. She has authored and co-authored chapters in books, articles in national and international journals, and has made various communications and lectures at international conferences and seminars. Email: aor@uevora.pt.

Cláudia Seabra, Ph.D., is Professor of Tourism and Coordinator of the PhD in Tourism, Heritage and Territories in the University of Coimbra, Portugal. Her research areas include tourism and safety and consumer behaviour. She has been published in several tourism, leisure and economics journals and has authored book chapters. She is theme editor of the *International Journal of Tourism Cities - Terrorism in Tourism Cities.* E-mail: cseabra@uc.pt.

Carla Silva, Ph.D., is Professor of Sociology of Tourism and Coordinator of the Master in Tourism Management at School of Technology and Management, Polytechnic Institute of Viseu (Portugal). Her research interests include culture of consumption, storytelling, mountain and nature-based tourism with special interest in tourism motivations, place attachment and tourism imagery. E-mail: csilva@estgv.ipv.pt.

Liyong Wang, Ph.D., is Associate Professor of International Marketing at Otaru University of Commerce, Japan. His research interests include global marketing, consumer behavior, and tourism marketing. E-mail: wanglyfrank@res.otaru-uc.ac.jp.

1 Introduction: Collaboration and technology for more sustainable and responsible tourism marketing

Antónia Correia and Alain Decrop

Vulnerable is how we are nowadays. In fact, the impact of the Covid-19 pandemic is neither time limited nor spatially contained. But like many other natural disasters, the pandemic brought calamities and inequalities (Shklar, 1990), threatens the environment and raises a problem of precarity that is no longer limited to the poor and dependent as contagion patterns have no boundaries (Forester & McKibbon, 2020). So even more than revealing the vulnerability wealthy countries are facing, the pandemic forces us to recognize our progressively more interdependent lives in a globalized world and the responsibility to safeguard the planet.

Economies all over the world were hindered by Covid-19 but tourism was completely devasted by this pandemic. In the first five months of 2020, international tourism arrivals decreased by more than half and some $320 billion dollars in exports from tourism were lost. Overall, some 120 million direct jobs in tourism are at risk (WTO, 2020). The current situation recalls emergency status for countries that depend on tourism and for minorities that may find in tourism a driver to social integration, empowerment and income. The Covid-19 crisis offers opportunities to rebuild tourism in a safe, equitable and sustainable way. To that end technology, partnerships and sustainable and responsible practices are strategic.

These strategic axes are not new; even in times when tourism was boosting economies all over the world, sharing practices, responsibility and sustainability were brought to the fore as the way to not perish together. Back to the fundamentals of economics, general equilibrium is today more than ever a priority. Learning to live together in a responsible and sustainable way is mandatory to cope with the vulnerability we are facing.

This book was planned in 2019 when the problems of tourism were about mass and over-growth. According to WTO (2017), tourism demand has tripled in the two last decades to reach 1.326 billion international tourist arrivals in 2017. Before the Covid-19 crisis, global tourism was expected to continue to grow in the future as a larger number of aspiring travelers become more prosperous and their disposable income increases. The sector was expected to outperform the global economy in 10 years, increasing by an estimated rate of 4% on average annually. Under this outperformance, environmental and societal impacts were outlined as strategic issues for further discussion.

This discussion started at the 8th ATMC (Advances in Tourism Marketing Conference) hosted by the University of Namur (NADI-Center for Research on Consumption and Leisure). Following the success of the previous conferences held throughout Europe, the 8th ATMC focused on marketing for more sustainable and collaborative tourism, seeking to address these two major challenges for the tourism industry. On the one hand, tourism marketing is more and more criticized as exploitative and fueling hedonistic consumerism. However, marketing skills and techniques can also be used for good purposes, by understanding market needs, designing more sustainable products and identifying more persuasive methods of communication to transform tourists' unsustainable behaviors. A series of papers included in the program presented theories, methods and results for enhancing such a more sustainable marketing. On the other hand, ATMC intended to focus on the sharing/collaborative economy, which is affecting tourism to a large extent. Collaborative business models, with champions such as Airbnb or Uber, are often presented as more sustainable than traditional ones, as they empower ordinary people and promote the shared use of resources. However, sharing practices in business raise new social challenges and ethical questions as well.

This book that developed around the best papers presented during the ATMC 2019 conference converges in a number of chapters that were by themselves a testimony to the power of collaboration in a network such as ATMC. In line with the themes outlined above, four key questions are addressed in the book: Can technology and connectivity contribute to value co-creation? How do sharing economy practices develop in tourism? How does sustainability impact on culture and the well-being of residents? How is overtourism affecting tourism sustainability and how are residents empowered to help places to be recognized as tourism destinations? To answer these questions, the book is organized in four parts where interdependence, collaboration, sustainability and responsible behavior are discussed in depth, with the aim of contributing to revamp tourism in and after the vulnerable times we are living now.

The first part, ***Technology and Value Co-Creation***, aims to discuss the level of interdependence needed to co-create value in tourism experiences. This part of the book comprises three chapters, which cover different types of connectivity to reach value. The first chapter by Michelle Moraes, Áurea Rodrigues, Antónia Correia and Metin Kozak, entitled '*Absorptive capacity, co-creation and tourism: a mixed analysis method*', assesses interdependences at macro level throughout a literature review that demonstrates companies' ability to innovate through absorption of new external information. Such an 'absorptive capacity' has become the focus of many studies in recent years, suggesting that innovation, co-creation and information are strategically intermingled in co-creation. The second chapter by Xing Han, Carolus Praet and Liyong Wang, '*Social interaction in co-creating the tourist experience: An exploratory study of Chinese visitors to Japan*', sheds light on how social interactions help to co-create tourism experiences, suggesting that Chinese tourists value interaction with residents, but not so much with other tourists. The third chapter by Nico Didry and Jean-Luc Giannelloni, '*Emotional interactions in festivals: How do consumers build a collective emotional experience?*', approaches one of the most Covid-impacted activities in tourism, i.e. festivals. Their work focuses on how festivalgoers establish emotional bounding and how they transfer such emotions to their in-groups. This chapter shows some paths to develop ambassadorships in festivals.

The second part, ***Platforms and Collaborative Economy***, is organized in three chapters exploring platforms within the sharing economy, such as EatWith, Airbnb and Couchsurfing, that all provide collaborative or 'shareable' tourism experiences. As in the first part, this set also starts with a macro approach of sharing economy principles in tourism experiences. The chapter by Silvana Canales Gutiérrez, '*Collaborative economy in the tourism industry: The new deal for consumers in the European Union*', enlists advantages and disadvantages of collaborative economy in tourism; first seen as an opportunity, this could also be very unfair to consumers if their rights are not respected. With this alert in mind, the chapter by Marina Petruzzi, Áurea Rodrigues, Michelle Moraes and Antónia Correia, '*An analysis of meal-sharing reviews to explore serendipity*', focuses on a very different tourism practice, i.e., sharing meals with locals, that most tourists do not know yet. Serendipity is presented as an enabler in such food sharing experiences that may be perceived as risky. The last chapter of this part by Marie Dewitte, Jérôme Mallargé and Alain Decrop, '*Consumer perception of service quality: The case of Airbnb and Couchsurfing*', investigates the service evaluation process in the case of peer-to-peer accommodation. Based on an analysis of online archival data and narratives, their study highlights that the dimensions taken into account in users' service quality assessment differ according to the type of (commercial vs. non-profit) platform.

The third part, ***Sustainable Tourism Development***, comprises two chapters that explore how tourists and hosts could interact to enable sustainable development and how cultural heritage could be developed in a more sustainable way. The chapter by Ali Ozturen, Arash Akhshik and Foad Irani, '*Host–tourist interactions and residents' attitudes towards sustainable tourism development*', aims to develop a theoretical model able to explain how tourists and residents may contribute to a better tourism development. In the next chapter entitled '*Challenges to sustainability in prospective World Heritage sites*', Sina Kuzuoglu and Stella Kladou investigate the cultural heritage resources of Iznik (Nicaea), a prospective World Heritage site in northwestern Turkey situated east to its namesake lake. More specifically, their research addresses how the site managers' and local officials' perspectives intertwine with the local community's interaction with heritage assets. The authors also show how administrative decision-making processes influence sustainability challenges, especially in their social and environmental components.

The fourth part of the book, ***Technology, Residents and Overtourism***, provides very practical examples of overtourism, and residents' and millennials' perceptions of tourism development. The chapter by Stéphane Bourliataux Lajoinie, Josep Lluis del Olmo Arriaga and Frédéric Dosquet, '*How Digital Strategy Increases Over-Tourism – The Case of Barcelona*', confronts digital strategies and tools with overcrowded destinations such as Barcelona, ending with a number of propositions in order to enhance value co-creation. The chapter by Giacomo Del Chiappa, Francesca Checchinato and Marcello Atzeni, '*Residents' perception of cruise tourists in an overcrowded city: The case of Venice*', extends the discussion on overtourism, this time through the perspective of local people. Finally, the chapter by Andreia Pereira, Carla Silva, Cláudia Seabra and Manuel Reis, *'Place attachment and residents' perceptions of tourism development in small town destinations'*, explores how residents perceive and accept tourism impacts in small towns.

Ending as we started, the conclusion section entitled *'Preparing for the future of travel & tourism in vulnerable times'*, by Alain Decrop and Antónia Correia, outlines the strategies presented in this book that could be adapted to the current pandemic situation.

References

Forester, J. & McKibbon, G. (2020). Beyond blame: leadership, collaboration and compassion in the time of COVID-19. *Socio-Ecological Practice Research*, 2(3), 205-216.

Shklar, J. (1990) *The Faces of Injustice.* Yale University Press, New Haven.

UNWTO World Tourism Organization. (2017). *Tourism Highlights, 2017 edition.*

UNWTO World Tourism Organization. (2020). *Tourism Highlights, 2020 edition.*

Part 1
Technology and Value (Co-)creation

2 Absorptive capacity, co-creation and tourism: A mixed analysis method

Michelle Moraes, Áurea Rodrigues, Antónia Correia and Metin Kozak

The objectives of this chapter are to:

- ☐ Elaborate a quantitative and qualitative analysis of the literature that combines absorptive capacity, co-creation and tourism.
- ☐ Identify research trends.
- ☐ Underline similarities and singularities that make it possible to converge research streams.

Keywords: absorptive capacity, co-creation, innovation

Introduction

Absorptive capacity is considered as an important innovation enabler since it can reduce risks and costs associated with innovations developed entirely by the companies themselves (Apriliyanti & Alon, 2007). Absorptive capacity can be understood as the ability to recognize the value of new external information and its application (Cohen & Levinthal, 1990). The number of publications about absorptive capacity have increased in recent years and, amongst the main streams studied in this field, there exist inter-organizational learning and knowledge transfer dynamics (Apriliyanti & Alon, 2007). In this context, the level of new external information availability was considered to be one of the main absorptive capacity enablers (Zahra & George, 2002). However, this availability is not only related to the amount of external information, but also to how this information is transferred (Markovic & Bagherzadeh, 2018). Consequently, co-creation has been highlighted as an important way to enhance knowledge sharing and to facilitate the absorption of new external information (Markovic & Bagherzadeh, 2018).

Concerning co-creation, it can be considered as a "new paradigm in the management literature" (Galvagno & Dalli, 2014: 643) and can be defined as a process that "involves the participation in the creation of the core offering itself. It can occur ... with customers and any other partners in the value network" (Lusch & Vargo, 2006: 284). Thus, it can be understood as a relationship between stakeholders (Azevedo, 2009) that emerged from active, creative and collaborative experiences (Campos et al., 2018; Kazadi, Lievens & Mahr, 2016) through which value is created (Haahti, 2006). Therefore, co-creation results in developing new products and services in a faster, more relevant and innovative way than traditional processes. It is a process that brings the opportunity to continue interaction between the firm and customers in which the firm is willing to work with external stakeholders and obtain more value through this collaboration with customers (Hamidi, Gharneh & Khajeheian, 2020). On the other hand, one of the ways to synthesize knowledge effectively is by developing the capacity for innovation, which has the virtue of establishing processes that take advantage of absorbed knowledge to incorporate new functions or create new products, and also to mitigate the impact of negative externalities, identify new sources of raw materials, access new services that add value, enter new underserved markets, improve operational processes, or adopt new activities for commercialization and business management (Rodríguez, Barón & Guaita, 2020).

In the tourism literature, co-creation has been approached through different perspectives, such as organization, destination and tourists (Campos et al., 2018). In the organizational context, competitive performance has been highlighted as a co-creation output; however, co-creation processes should be more focused (Campos et al., 2018). Absorptive capacity and co-creation are especially relevant to the tourism industry given its high dependence on external knowledge as an innovation source (Thomas & Wood, 2014) and its low performance profile in terms of:

i) all kind of innovation outputs – product, process, organizational and marketing (Camison & Monfort-Mir, 2012);
ii) level of cooperation (Pinto, Cruz & Combe, 2015) – including companies of the same group, customers and universities – in order to develop innovations (Camison & Monfort-Mir, 2012);
iii) non-academic impact – low absorption by organizations – of academic outputs (Thomas & Ormerod, 2017).

Tussyadiah and Zach (2014) suggest that the capacity to work together with stakeholders in order to innovate depends on the capacity to acquire knowledge, transform knowledge within their organizations and explore it for new service/product development. Thus, more studies are necessary to analyze and create a better understanding of these two constructs: absorptive capacity and co-creation in tourism innovation. With this in mind, the main objective of this chapter is to indicate, through a comparison with the configuration presented by absorptive capacity general literature, the boundaries of the research that also contemplated tourism and co-creation, and consequently, new paths to tourism research in this context.

Methodology

To elaborate a bibliometric analysis, we considered the WoS Core Collection management publications that had topics as 'absorptive capacity' and 'tourism' (30 studies), as well as those that included 'absorptive capacity' and 'co-creat*' or 'cocreat*' or 'co creat*' (55 works). The delimitation of WoS as a database is the fact that most cited bibliographic reviews about absorptive capacity contemplated only this collection. In order to obtain a more focused and a deeper analysis within this frame, it was decided to use 'management' as the field category of WoS Core Collection.

To analyze these studies, we focused on a mixed analysis method:

i) a bibliometric analysis of the terms of the publications that had 'absorptive capacity' as well as 'absorptive capacity' and 'tourism' or 'co-creation'as topics; and

ii) a qualitative analysis to verify the alignment between these publications ('absorptive capacity' vs. 'absorptive capacity' and 'tourism' or 'co-creation').

The source quantification (authors and publication names) of 'absorptive capacity' and 'co-creation' or 'tourism' literature was elaborated by using HistCite (12.03.17).

Regarding the qualitative analysis, all 85 WoS Core Collection management publications that had topics as 'absorptive capacity' and 'co-creation' or 'tourism' were fully analyzed with content analysis. Content analysis is described as a method to review texts and other data forms in a way to obtain relevant data from a thorough review of artifacts (Leavy, 2007). This method relates to systematic literature review approaches involving a more orderly and consistent method to map, consolidate and identify gaps in an existing body of knowledge (Gosling et al., 2016). In this study, we adopted the qualitative approach to develop the content-based analysis. Seuring and Gold (2012) describe the main steps involved in this method as follows:

i) **material collection** – delimitation of the material and unit of analysis;

ii) **descriptive analysis** – initial descriptive analysis of the material;

iii) **category selection** – selection of the collected material according to specific analytic categories or dimensions that considered ACAP traditional categories present in Gao et al. (2017), which is the Web of Science paper in English that contemplates in its title the words 'absorptive capacity' and 'literature analyses' (Figure 2.1); and

iv) **material evaluation** – theoretically based analysis of the material according to the categories previously specified.

This selective approach provides a helpful methodological basis for the examination of research work in a systematic way, allowing convergence of focus only on the works considered most significant and relevant to the theoretical aspects being reviewed. This analysis was done using NVivo (12) which is a software package to digitally code texts or images, that allow the user to synchronize evidence and make analytically richer intersections (Hai-Jew, 2017). By using this software, the researcher is able to manage the empirical material in a single location (text, spreadsheets, audio, video or images) and can add, modify, connect and cross reference data, or even record ideas in the form of memos in order not to lose any insights generated while viewing the material (Edhlund & McDougall, 2019).

Figure 2.1: Absorptive capacity process traditional perspectives, model components.

		Antecedents						Components						
Paper	Year	Knowledge Source / Complementarity / Interorganizational Antecedents	Prior Knowledge	Intraorganizational Antecedents	Managerial Antecedents	Learning Relationships / Individual Development	Environmental Conditions / Incentives	PACAP	RACAP	Recognizing the Value	Acquire	Assimilate	Transform	Apply / Exploit
Cohen and Levinthal	1990	X	X							X		X		X
Zahra and George	2002	X	X					X	X		X	X	X	X
Lane et al.	2006	X	X			X	X			X		X		X
Todorova and Durisin	2007	X	X							X	X	X	X	X
Volberda et al.	2010	X	X	X	X	X	X	X	X		X	X	X	X

		Contingent Factors								Outcomes							
Paper	Year	Regimes of Appropriability	Activation Triggers	Social Integration Mechanisms	Environmental Conditions	Organisational Mental Models	Organisational Strategies	Organisational Structures and Processes	Power Relationships	Innovation	Innovative Performance	Knowledge Outputs	Exploitation / Exploration	Commercial Outputs	Flexibility	Performance	Competitive Advantage
Cohen and Levinthal	1990	X								X	X						
Zahra and George	2002	X	X	X						X					X	X	X
Lane et al.	2006					X	X					X		X		X	
Todorova and Durisin	2007	X	X	X				X		X					X	X	X
Volberda et al.	2010	X			X				X	X			X			X	X

Source: Gao et al. (2017)

According to Gao et al. (2017), the aspects of absorptive capacity present in some of the most quoted traditional models can be divided into absorptive capacity antecedents, components, contingent factors and outcomes. Regarding 'antecedents', this is mainly composed by aspects such as:

i) availability/variety of external knowledge sources;
ii) partner complementarity;
iii) previous experience in collaboration;
iv) prior knowledge basis;
v) management cognition;
vi) establishment of learning relationships within networks; and
vii) competitiveness level.

Concerning 'components', some of them are acquisition, assimilation, transformation and application of new external knowledge (Zahra & George, 2002). These stages are related, respectively, to identification, incorporation and internal changings needed to use (application) new external knowledge (Zahra & George, 2002). Some authors aggregate these categories into two groups and consider that acquisition and assimilation are related to potential absorptive capacity – PACAP, while transformation and application are associated to realized absorptive capacity – RACAP (Zahra & George, 2002). 'Contingent factors' are moderators in organizational structures and strategies, to draw relationships and intellectual property (Gao et al., 2017). As for the absorptive capacity 'outcomes', some of the main aspects are innovation, performance and competitive advantage (Gao et al., 2017).

Results

As mentioned earlier, the quantitative and qualitative analysis was centered in the 85 WoS Core Collection management publications that had topics as 'absorptive capacity' and 'co-creation' or 'tourism' recurring at HistCite, and content analysis with NVivo. A mixed method approach was chosen while the "diversity of methods implies rich opportunities for cross-validating and cross-fertilizing research procedures, findings, and theories" (Brewer & Hunter, 2006: 1).

Quantitative analysis

Regarding the number of Web of Science management publications about absorptive capacity (ACAP), it has increased in the past years (refer to Figure 2.2). This tendency was also confirmed in the absorptive capacity management literature that contemplated tourism or co-creation dimensions in their main topics. However, there are none of WoS management publications that simultaneously contemplate in their topics the terms 'absorptive capacity' and 'tourism' and 'co-creation.' In 2018, the number of Web of Science publications with 'absorptive capacity' (ACAP) in their topics was 390. The total number of works that additionally included 'tourism' (TOUR) or 'co-creation' (CC) was 7 and 11, respectively.

Figure 2.2: Number of Web of Science management publications about absorptive capacity (and tourism and/or co-creation) between 1996 and 2018

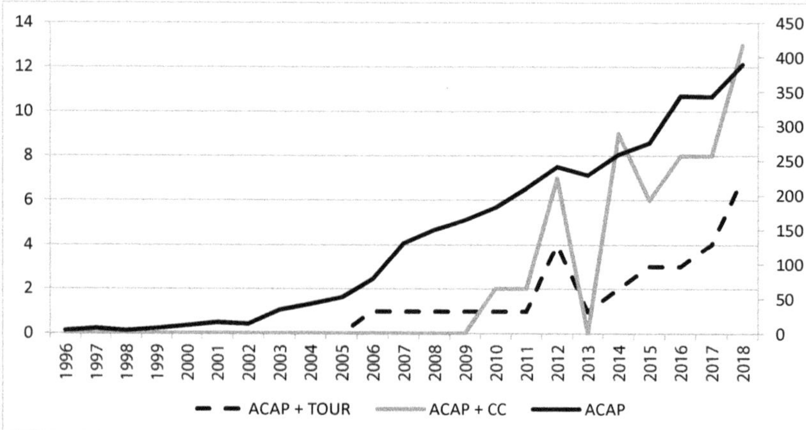

Source: Authors' elaboration.

In terms of the number of publications by source, the distribution of the publications that contemplated the terms 'absorptive capacity' and 'tourism' was more heterogeneous than other literature – 'absorptive capacity' and 'co-creation' (refer to Table 2.1). In the first case, the journals with the highest numbers of publications were *Tourism Management* and *International Journal of Contemporary Hospitality Management* with 14 and 4 studies, respectively.

Table 2.1: Number of publications by source (more than two publications)

ACAP + TOUR	N° Pub	% Pub	TLCS	TGCS	ACAP + CC	N° Pub	% Pub	TLCS	TGCS
Tourism Management	14	47%	45	593	*Journal of Product Innovation Management*	6	11%	7	603
International Journal of Contemporary Hospitality Management	4	13%	4	78	*Technological Forecasting and Social Change*	5	9%	1	30
					Industrial Marketing Management	3	5%	1	69
					Journal of Business Research			3	63

Source: Authors' elaboration.

On the other hand, the sources with more than two works about 'absorptive capacity' and 'co-creation' were *Journal of Product Innovation Management* (6), *Technological Forecasting Social Change* (5), *Industrial Marketing Management* (3) and *Journal of Business Research* (3). Regarding the highest number of total local and global citation scores, *Tourism Management* and *Journal of Product Innovation Management* achieved the highest, respectively, in the literature that contemplated tourism and co-creation as well.

The distribution of these publications by document type was very heterogeneous in both cases (Table 2.2). 'Article' was the most recurrent category in both. 'Articles' also had the highest total local (TLCS) and global citation (TGCS) scores in the publications related to 'absorptive capacity' and 'tourism.' However, 'Review' achieved the maximum global citation scores in the literature associated with 'absorptive capacity' and 'co-creation.' In this literature that also contemplated co-creation, 'article' and 'review' were the categories that presented the topmost total local citation scores.

Table 2.2: Number of publications by document type

ACAP + TOUR	N° Pub	% Pub	TLCS	TGCS	ACAP + CC	N° Pub	% Pub	TLCS	TGCS
Article	25	83%	36	570	Article	45	82%	8	560
Proceedings Paper	3	10%	0	0	Review	7	12.7%	8	668
Review	2	7%	13	179	Proceedings Paper	2	3,6%	0	1
					Editorial Material	1	2%	0	11

Source: Authors' elaboration.

Regarding the number of works by author, Thomas (Thomas, 2012; Thomas & Ormerod, 2017; Thomas & Wood, 2014; Thomas & Wood, 2015;) and Nieves (Nieves & Diaz-Meneses, 2018; Nieves & Haller, 2014) had the highest number of studies that contemplated the terms 'absorptive capacity' and 'tourism' of 4 and 2 studies, respectively (Table 2.3). Their TLCS were 14 and 6, respectively. However, the peak in terms of TGCS in this literature was achieved by Shaw and Williams (TGCS = 179) that had only one publication – together – in this field and, for this reason, they do not appear in Table 2.3. In case of publications related to 'absorptive capacity' and 'co-creation', Wilden (Randhawa, Wilden & Gudergan, 2018; Randhawa, Wilden & Hohberger, 2016; Wilden et al., 2019) was the single author with more than two publications. However, the greatest global citation score was not achieved by him, but by West and Bogers

(2014) with specifically more than 417 citations that had only one study that contemplated these dimensions as well.

Table 2.3: Number of publications by author (more than 1 publication)

<table>
<tr><th>ACAP + TOUR</th><th>N° Pub</th><th>TLCS</th><th>TGCS</th><th>ACAP + CC</th><th>N° Pub</th><th>TLCS</th><th>TGCS</th></tr>
<tr><td>Thomas R</td><td>4</td><td>14</td><td>111</td><td>Wilden R</td><td>3</td><td>3</td><td>111</td></tr>
<tr><td>Nieves J</td><td>2</td><td>6</td><td>76</td><td>Brandl K</td><td rowspan="11">2</td><td>1</td><td>10</td></tr>
<tr><td></td><td></td><td></td><td></td><td>Chang KH</td><td>0</td><td>41</td></tr>
<tr><td></td><td></td><td></td><td></td><td>Chen CJ</td><td>0</td><td>8</td></tr>
<tr><td></td><td></td><td></td><td></td><td>Gudergan S</td><td>0</td><td>9</td></tr>
<tr><td></td><td></td><td></td><td></td><td>Lin BW</td><td>0</td><td>8</td></tr>
<tr><td></td><td></td><td></td><td></td><td>Morgan T</td><td>1</td><td>12</td></tr>
<tr><td></td><td></td><td></td><td></td><td>Randhawa K</td><td>3</td><td>108</td></tr>
<tr><td></td><td></td><td></td><td></td><td>Sarker S</td><td>0</td><td>186</td></tr>
<tr><td></td><td></td><td></td><td></td><td>Su CY</td><td>0</td><td>8</td></tr>
<tr><td></td><td></td><td></td><td></td><td>Van Geenhuizen M</td><td>1</td><td>16</td></tr>
</table>

Source: Authors' elaboration.

Concerning distribution by country, UK and USA were the top two origins (Table 2.4). Spain and Finland were respectively on this list of studies that contemplated 'tourism' and 'co-creation'. In terms of total local and global citation scores, UK held the best results in literature that contemplated tourism, while USA achieved that in relation to co-creation. This configuration demonstrates the strength of these regions in this field, in terms of quantity and quality.

Table 2.4: Number of publications by country (top 3)

<table>
<tr><th>ACAP and TOUR</th><th>N° Pub</th><th>% Pub</th><th>TLCS</th><th>TGCS</th><th>ACAP and CC</th><th>N° Pub</th><th>% Pub</th><th>TLCS</th><th>TGCS</th></tr>
<tr><td>UK</td><td>10</td><td>33%</td><td>28</td><td>313</td><td>UK</td><td rowspan="2">10</td><td rowspan="2">18%</td><td>2</td><td>129</td></tr>
<tr><td>Spain</td><td>6</td><td>20%</td><td>16</td><td>285</td><td>USA</td><td>6</td><td>657</td></tr>
<tr><td>USA</td><td>4</td><td>13%</td><td>0</td><td>57</td><td>Finland</td><td>8</td><td>15%</td><td>0</td><td>86</td></tr>
</table>

Source: Authors' elaboration.

Absorptive capacity and tourism

Different approaches were used to analyze absorptive capacity within the tourism context (Binder, 2019). In this paper, absorptive capacity has generally appeared together with dimensions such as knowledge (transfer), cooperation, learning and dynamic capabilities (Camison & Monfort-Mir, 2012; Omerzel, 2016; Murray, Lynch & Foley, 2016). In this part of the chapter, the studies that contemplated simultaneously 'absorptive capacity' and 'tourism' were analyzed through Gao et al. (2017) with absorptive capacity dimensions mentioned previously, and more specifically ACAP antecedents, components, contingent factors and outcomes.

Amongst the ACAP antecedents in the 'absorptive capacity' and 'tourism' literature, there were traditional aspects, such as :

i) prior knowledge and skills (Nieves & Haller, 2014);

ii) interorganizational cooperation (Wilke et al., 2019);

iii) cooperative-competitive tension in knowledge sharing between hotels (Idrees, Vasconcelos & Ellis, 2018);

iv) existence of knowledge where companies were situated, including those that came from universities and affiliation to a hotel chain (Marco-Lajara et al., 2018).

A specific antecedent associated with the tourism industry was reciprocity (Thomas & Wood, 2015) and personalized source of knowledge (Thomas & Wood, 2014).

The ACAP components designed by Zahra and George (2002) and Cohen and Levinthal (1990), or with little variation of both, were recurrent in this literature (Elbaz, Agag & Alkathiri, 2018; Garay, Font & Pereira-Moliner, 2017; Hoarau, 2014; Otengei et al., 2017; Pinto, Cruz & Combe, 2015; Thomas & Wood, 2014, 2015; Zavattaro, Daspit & Adams, 2015; Wilke et al., 2019). Tourism companies' knowledge acquisition processes were more associated with industry-related sources (Garay, Font & Pereira-Moliner, 2017) and were dependent on the type of external knowledge sources: learning from competitors, for example, occurred through direct observation, while from customers it was associated with interaction, feedback and questionnaires (Hoarau, 2014). In the case of the assimilation step, the relevance of some tacit and explicit aspects was highlighted, such as:

i) sharing tacit knowledge with colleagues;

ii) management accessing tacit knowledge; and

iii) spreading knowledge among employees through guiding handbooks and newsletters (Hoarau, 2014).

Moreover, in the assimilation step, there was a "dissociation between innovations adopted by directly incorporating the specific knowledge provided by external agents and innovations that require the mediation of intra-organizational collaboration for their development" (Nieves & Diaz-Meneses, 2018: 2537).

In terms of ACAP components' proxies, they were considered traditional and innovative aspects. In the first group, there were dimensions such as "the ratio of graduates over technicians" (Kumar, Kumar & de Grosbois, 2008: 19) and employees' absorptive capacity, more specifically employees' capacity to identify, process and apply new knowledge to commercial ends (Elbaz, Agag & Alkathiri, 2018). Amongst the nontraditional dimensions, there was personnel expenses per room (Marco-Lajara et al., 2018). However, the relevance of R&D expenses, a central aspect in the traditional ACAP literature, was not considered a proxy in these studies, especially in the case of small-size companies (Marco-Lajara et al., 2018).

Some traditional contingent factors also appeared in this literature: the level of informalization, leadership and its proximity to core knowledge employees and market, as well as tacit knowledge obtained from experience in the industry (Thomas & Wood, 2015). Apropos the regimes of appropriability, in the tourism context, it was considered appropriated only in case of innovation with significant technological dimension (Thomas & Wood, 2015). Concerning the knowledge transferring dynamics within the tourism industry related to workers' movements, it was considered an important, but an understudied subject (Shaw & Williams, 2009). In terms of the general profile of employment in the tourism industry – seasonal, high level of turnover and inadequately qualified – it was emphasized that it did not contribute to tourism organizations' ACAP (Cooper, 2015).

Concerning the outcomes, they were also related to traditional results, such as improvement of performance (Elbaz, Agag & Alkathiri, 2018; Garay, Font & Pereira-Moliner, 2017), competitive advantage (Wilke et al., 2019), innovation and flexibility (Thomas & Wood, 2014) through aspects of:

i) development of existing offerings (Otengei et al., 2017; Murray, Lynch & Foley, 2016; Zavattaro, Daspit & Adams, 2015) and marketing campaigns (Zavattaro, Daspit & Adams, 2015);
ii) delay reduction in terms of scientific and technological progress adoption (Hjalager, 2015);
iii) moderation between explorative learning and knowledge transfer (Liu, 2018);
iv) enhancement in terms of joining benefits from foreign investment externalities (Mao & Yang, 2016);
v) being green innovators (Pace, 2016); and
vi) establishment of collaboration (Pinto, Cruz & Combe, 2015).

A model of absorptive capacity in tourism inspired by the considered literature is presented in Figure 2.3, through which it is possible to verify that the most part of ACAP dimensions present in this literature is contemplated by the ACAP traditional models (refer to Figure 2.1). Concerning the innovative aspects (*) understood as singularities of this literature and industry, they are mainly related to ACAP antecedents (reciprocity and personalized source of knowledge) and contingent factors (regimes of appropriability in the case of innovation with significant technological dimension and general profile of employment). As co-creation is an important aspect to knowledge transfer processes, it is also important to analyze the literature associating it with absorptive capacity.

Figure 2.3: Absorptive capacity and tourism model

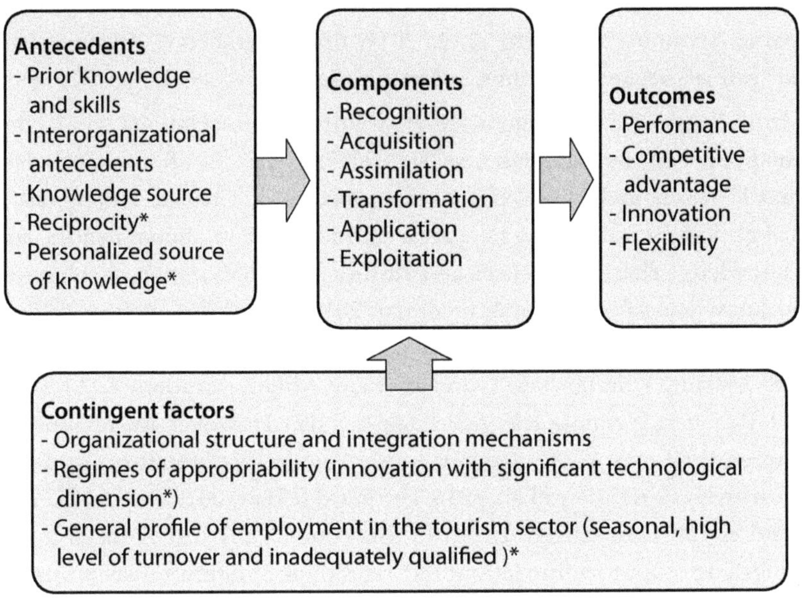

Source: Authors' elaboration.

Absorptive capacity and co-creation

The general idea in this study is that benefits from interaction with stakeholders in value co-creation processes also depends on organizations' ACAP level (Mäkinen, Kanniainen & Peltola, 2014; Morgan, Obal & Anokhin, 2018; Paswan, D'Souza & Rajamma, 2014; Wilden et al., 2019), including in the case of SMEs (Del Giudice et al., 2019) and universities (Miller, Mcadam & Mcadam, 2014). In this context, market orientation is not enough for the achievement of high performance since how the stakeholders were involved was considered fundamental (Medhi, Jain & Jain, 2019; Morgan, Anokhin & Wincent, 2019; Sarker et al., 2012).

Regarding ACAP antecedents, this literature contemplated traditional aspects of intra/interorganizational characteristics, for instance, stakeholders' characteristics (De Silva & Rossi, 2018; Komulainen, 2014; Marwede & Herstatt, 2019; Mol & Brandl, 2018) and knowledge network configuration (Guimón & Paraskevopoulou, 2017; Miller, Mcadam & Mcadam, 2014; Pucci et al., 2018; Su, Lin & Chen, 2016; Van Geenhuizen & Nijkamp, 2012). New enablers also emerged from this context as in the case of stakeholders' sense-making and sense-giving (Prior, Keränen & Koskela, 2018). Similarly, in the literature on tourism, ACAP traditional phases designed by Zahra and George (2002) and Cohen and Levinthal (1990) were present in this literature (Chew, 2012; Hakanen, 2014; Jane & Wang, 2012, Morgan, Obal & Anokhin, 2018; Pittz et al., 2019; Prior, Keränen & Koskela, 2018) and were associated with innovation integration (West & Bogers, 2014).

In the case of a contingent factor, traditional associations were established between routinization and ACAP (Pittz et al., 2019) as well as dedicated IT assets and knowledge exchange (Chang, Chen & Huang, 2015). Furthermore, it ratified the relevance of traditional relational factors, such as knowledge sharing practices and attitudes to collaboration development and knowledge flow (Chai & Freeman, 2019; Chang & Gotcher, 2010; De Silva & Rossi, 2018; Mariano & Awazu, 2017; Markovic & Bagherzadeh, 2018; Martín, Reinhardt & Gurtner, 2016; Miller, Mcadam & Mcadam, 2014; Parjanen, Hennala & Konsti-Laakso, 2012). However, the relationship between trust and ACAP was not consensual in this literature (Anderson & Hardwick, 2017; Lew et al., 2016; Peronard & Brix, 2019; Pittz et al., 2019; Taheri & van Geenhuizen, 2016). In this context, the use of social media, connected to both routinization and relational dynamics, was an important factor inherent in the interaction between customers and companies (Bhimani, Mention & Barlatier, 2019). Another contingent factor confirmed by this literature was creativity promotion culture (Belkahla & Triki, 2011).

Concerning ACAP outputs, some of them were:

i) business value enhancement (Hakanen, 2014);

ii) internationalization of performance improvement (Eerme & Nummela, 2019);

iii) better performances in terms of specific types of new product development (Morgan, Obal & Anokhin, 2018; Su, Lin & Chen, 2016);

iv) knowledge creation (Kazadi, Lievens & Mahr, 2016).

A characteristic of this literature that is also present in the traditional ACAP publications was the concern with lock-in effects of co-creation processes: "by integrating customers at greater depth, we report that firms that are highly market-oriented may be less able to focus on integrating resources and transfer tacit knowledge from alternative customer groups and stakeholders, thus decreasing the firm's overall new product innovation performance" (Morgan, Anokhin & Wincent, 2019: 1115).

Through the model of absorptive capacity and co-creation based on the considered literature (Figure 2.4), it is possible to ratify the relevance of ACAP traditional dimensions to co-creation processes. In terms of singularities of this literature, some aspects that emerged were stakeholders' sense-making and sense-giving, as well as the use of social media and knowledge creation.

Figure 2.4: Absorptive capacity and co-creation model

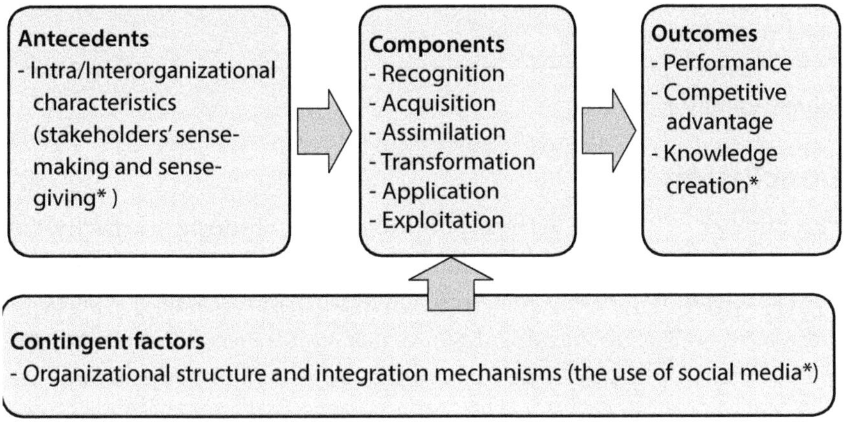

Source: Authors' elaboration.

A theoretical model of absorptive capacity and co-creation and tourism is presented (Figure 2.5). It highlights the traditional aspects and the singularities that emerged in the considered literature. Amongst the main considerations that can be extracted from this model are the secondary role of R&D expenses and regimes of appropriability – central aspects in the traditional ACAP literature, the relevance attributed to relational aspects and to general profile of employment in the tourism sector.

Figure 2.5: A model of absorptive capacity, co-creation and tourism

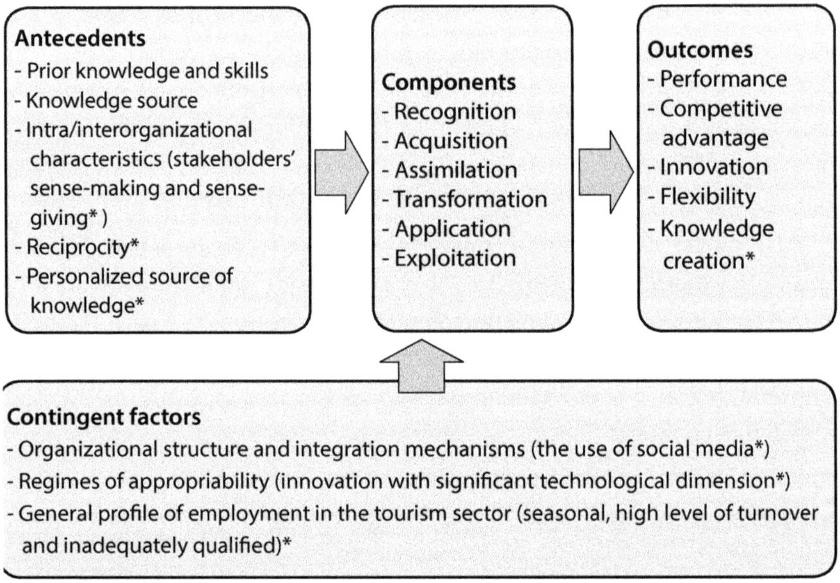

Source: Authors' elaboration.

Conclusion

The number of Web of Science publications on absorptive capacity has increased in recent years. This tendency was also confirmed in the absorptive capacity literature that contemplated tourism or co-creation aspects. As there were no WoS management works that simultaneously contemplated in their main topics the words 'absorptive capacity' and 'tourism' and 'co-creation', this study had to analyze these works separately, and more specifically: 'absorptive capacity' and 'tourism' and 'absorptive capacity' and 'co-creation'. The main results that emerged highlighted the strong alignment between the configuration present in the general absorptive capacity management literature and the two other specific branches ('tourism' or 'co-creation'). Substantially, the ACAP phases and outputs developed by the main traditional models, for example, were used. However, R&D, one of the key aspects of general ACAP publications, was not ratified by these two literatures. Furthermore, some important specificities were present in the publications that also contemplated tourism, such as:

i) antecedents – reciprocity and personalized source of knowledge;
ii) ACAP components' proxies – personnel expenses per room; and
iii) contingent factors – the general profile of employment in the tourism sector.

Concerning the studies that considered co-creation as well, some of the singularities were in terms of antecedents (stakeholders' sense-making and sense-giving), contingent factors (the use of social media) and outcomes (knowledge creation).

The main contribution of this chapter was the comparative analysis between the general absorptive capacity literature and the publications that also contemplated tourism and co-creation dynamics. There are few studies that explored the tourism companies' absorptive capacity in the process of co-creation, but neither on the singularities of absorptive capacity in the tourism industry nor on the co-creation process. Future studies should focus on how to open the tourism industry to the absorptive capacity black box (Thomas & Wood, 2015) in co-creation experiences while involving all the stakeholders – including universities. Consequently, possible starting points could be the analysis on the relationships between the singularities uncovered by this chapter. Furthermore, organizational aspects inherent in the co-creation process in the tourism industry that have already been studied, such as the relevance of information technology readiness (Cabiddu, Lui & Piccoli, 2013) should also be contemplated by future research. The main limitation of this study is relevant to the restriction of the terms used to extract the publications from the Web of Science database.

References

Apriliyanti, I. D. & Alon, I. (2007) 'Bibliometric analysis of absorptive capacity', *International Business Review*, 26 (5), 896-907.

Anderson, A. R. & Hardwick, J. (2017) 'Collaborating for innovation: The socialised management of knowledge', *International Entrepreneurship and Management Journal*, 13 (4), 1181–1197.

Azevedo, A. (2009) 'Designing unique and memorable experiences: Co-creation and the "surprise" factor.' paper presented at the III Congresso Internacional de Turismo de Leiria e Oeste.

Belkahla, W. & Triki, A. (2011) 'Customer knowledge enabled innovation capability: Proposing a measurement scale', *Journal of Knowledge Management*, 15 (4), 648–674.

Bhimani, H., Mention, A. L. & Barlatier, P. J. (2019) 'Social media and innovation: A systematic literature review and future research directions', *Technological Forecasting and Social Change*. 144, 251–269.

Binder, P. (2019) 'A network perspective on organizational learning research in tourism and hospitality: A systematic literature review', *International Journal of Contemporary Hospitality Management*, 31 (7), 2602–2625.

Brewer, J. & Hunter, A. (2006) *Foundations of Multimethod Research: Synthesizing styles*. Thousand Oaks, CA: Sage.

Cabiddu, F., Lui, T.-W. and Piccoli, G. (2013) 'Managing value co-creation in the tourism industry', *Annals of Tourism Research*, 42, 86-107.

Camison, C. & Monfort-Mir, V. M. (2012) 'Measuring innovation in tourism from the Schumpeterian and the dynamic-capabilities perspectives', *Tourism Management*, 33 (4), 776-789.

Campos, A. C., Mendes, J., Valle, P. O. & Scott, N. (2018) 'Co-creation of tourist experiences: a literature review', *Current Issues in Tourism*, 21 (4), 369-400.

Chai, S. & Freeman, R. B. (2019) 'Temporary colocation and collaborative discovery: Who confers at conferences', *Strategic Management Journal*, 40 (13), 2138–2164.

Chang, K. H. & Gotcher, D. F. (2010) 'Conflict-coordination learning in marketing channel relationships: The distributor view', *Industrial Marketing Management*, 39 (2), 287–297.

Chang, K. H., Chen, Y. & Huang, H. F. (2015) 'Information technology and partnership dynamic capabilities in international subcontracting relationships', *International Business Review*, 24 (2), 276–286.

Chew, E. (2012) 'Sustainable value creation through entrepreneurial leadership in SME' paper presented at the 8th European Conference on Management Leadership and Governance.

Cohen, W. M. & Levinthal, D. A. (1990) 'Absorptive capacity: A new perspective on learning and innovation', *Administrative Science Quarterly*, 35 (1), 128-152.

Cooper, C. (2015) 'Managing tourism knowledge', *Tourism Recreation Research*, 40 (1), 107-119.

De Silva, M. & Rossi, F. (2018) 'The effect of firms' relational capabilities on knowledge acquisition and co-creation with universities', *Technological Forecasting and Social Change*, 133, 72–84.

Del Giudice, M., Scuotto, V., Garcia-Perez, A. & Messeni Petruzzelli, A. (2019) 'Shifting Wealth II in Chinese economy. The effect of the horizontal technology spillover for SMEs for international growth', *Technological Forecasting and Social Change*, 145, 307–316.

Edhlund, B. & McDougall, A. (2019) *NVivo 12 Essentials*. Stallarholmen, Sweden: Form & Kunskap.

Eerme, T. & Nummela, N. (2019), 'Capitalising on knowledge from big-science centres for internationalisation', *International Marketing Review*, 36 (1), 108-130.

Elbaz, A. M., Agag, G. M. & Alkathiri, N. A. (2018) 'How ability, motivation and opportunity influence travel agents' performance: The moderating role of absorptive capacity', *Journal of Knowledge Management*, 22 (1), 119–141.

Gao, S., Yeoh, W., Wong, S. F. & Scheepers, R. (2017) 'A literature analysis of the use of Absorptive Capacity construct in IS research', *International Journal of Information Management*, 37 (2), 36-42.

Galvagno, M. & Dalli, D. (2014) 'Theory of value co-creation: A systematic literature review', *Managing Service Quality*, 24 (6), 643–683.

Garay, L., Font, X. & Pereira-Moliner, J. (2017) 'Understanding sustainability behaviour: The relationship between information acquisition, proactivity and performance', *Tourism Management*, 60, 418–429.

Gosling, J., Jia, F., Yu, G. & Brown, S. (2016) 'The role of supply chain leadership in the learning of sustainable practice: Toward an integrated framework', *Journal of Cleaner Production*, 137, 1458–1469.

Guimón, J. & Paraskevopoulou, E. (2017) 'Factors shaping the international knowledge connectivity of industrial clusters: a comparative study of two Latin American cases', *Entrepreneurship and Regional Development*, 29 (9–10), 817–846.

Hakanen, T. (2014) 'Co-creating integrated solutions within business networks: The KAM team as knowledge integrator', *Industrial Marketing Management*, 43 (7), 1195–1203.

Haahti, A. (2006) 'Experience design management as creation of identity economies: Reflections from periphery on entrepreneurial designs in tourism' paper presented at the International Conference of Bulgarian Association for Management Development and Entrepreneurship.

Hai-Jew, S. (2017). *Social Media Data Extraction and Content Analysis*. Hershey, PA, USA: IGI Global.

Hjalager, A. M. (2015) '100 innovations that transformed tourism', *Journal of Travel Research*, 54 (1), 3-21.

Hamidi, F., Gharneh, N. & Khajeheian, D. (2020) 'A conceptual framework of value co-creation in service enterprises (case of tourism agencies)', *Sustainability*, 12, 213-224.

Hoarau, H. (2014) 'Knowledge acquisition and assimilation in tourism-innovation processes', *Scandinavian Journal of Hospitality and Tourism*, 14 (2), 135-151.

Idrees, I., Vasconcelos, A. & Ellis, D. (2018), 'Clique and elite: inter-organizational knowledge sharing across five star hotels in the Saudi Arabian religious tourism and hospitality industry', *Journal of Knowledge Management*, 22 (6), 1358-1378.

Jane, L.T. & Wang, T-F. (2012) 'Technology-based new service development process of small-enterprises' paper presented at the PICMET '12: Technology Management for Emerging Technologies, Vancouver.

Kazadi, K., Lievens, A. & Mahr, D. (2016) 'Stakeholder co-creation during the innovation process: Identifying capabilities for knowledge creation among multiple stakeholders', *Journal of Business Research*, 69 (2), 525–540.

Komulainen, H. (2014) 'The role of learning in value co-creation in new technological B2B services', *Journal of Business & Industrial Marketing*, 29 (3), 238-252.

Kumar, U., Kumar, V. & de Grosbois, D. (2008) 'Development of technological capability by Cuban hospitality organizations', *International Journal of Hospitality Management*, 27 (1), 12–22.

Leavy, P. (2007) 'The feminist practice of content analysis' In S. N. Hesse-Biber & P. L. Leavy (eds.), *Feminist Research Practice: A Primer*. Thousand Oaks, California: Sage Publications.

Lew, Y. K., Sinkovics, R. R., Yamin, M. & Khan, Z. (2016) 'Trans-specialization understanding in international technology alliances: The influence of cultural distance', *Journal of International Business Studies*, 47 (5), 577–594.

Liu, C. H. S. (2018) 'Examining social capital, organizational learning and knowledge transfer in cultural and creative industries of practice', *Tourism Management*, 64, 258-270.

Lusch, R. F. & Vargo, S. L. (2006) 'Service-dominant logic: reactions, reflections and refinements', *Marketing Theory*, 6 (3), 281–288.

Mäkinen S. J., Kanniainen J. & Peltola, I. (2014) 'Investigating Adoption of Free Beta Applications in a Platform-Based Business Ecosystem', *Journal of Product Innovation Management*, 31 (3), 451-465.

Mao, Z. X. & Yang, Y. (2016) 'FDI spillovers in the Chinese hotel industry: The role of geographic regions, star-rating classifications, ownership types, and foreign capital origins', *Tourism Management*, 54, 1-12.

Marco-Lajara, B., Zaragoza-Saez, P. D., Claver-Cortes, E. & Ubeda-Garcia, M. (2018) 'Hotel performance and knowledge sources in Spanish tourism districts', *Current Issues in Tourism*, 21 (17), 1988-2013.

Mariano, S. & Awazu, Y. (2017), 'The role of collaborative knowledge building in the co-creation of artifacts: influencing factors and propositions', *Journal of Knowledge Management*, 21 (4), 779-795.

Markovic, S. & Bagherzadeh, M. (2018) 'How does breadth of external stakeholder co-creation influence innovation performance? Analyzing the mediating roles of knowledge sharing and product innovation', *Journal of Business Research*, 88, 173-186.

Martín, M. V., Reinhardt, R. & Gurtner, S. (2016) 'Stakeholder integration in new product development: a systematic analysis of drivers and firm capabilities', *R&D Management*, 46 (S3), 1095–1112.

Marwede, M. & Herstatt, C. (2019) 'No innovation for the elderly? The influence of cognitive distance in corporate innovation', *Creativity and Innovation Management*, 28 (3), 355–367.

Medhi, P., Jain, P. & Jain, T. (2019), 'Effects of information sources for new customers and suppliers on the immediate innovation output of firms', *European Journal of Innovation Management*, 22 (4), 660-680.

Miller, K., Mcadam, M. & Mcadam, R. (2014) 'The changing university business model: A stakeholder perspective', *R&D Management*, 44 (3), 265–287.

Mol, M. J. & Brandl, K. (2018) 'Bridging what we know: The effect of cognitive distance on knowledge-intensive business services produced offshore', *International Business Review*, 27 (3), 669–677.

Morgan, T., Anokhin, S. A. & Wincent, J. (2019) 'Influence of market orientation on performance: the moderating roles of customer participation breadth and depth in new product development', *Industry and Innovation*, 26 (9), 1103–1120.

Morgan, T., Obal, M. & Anokhin, S. (2018) 'Customer participation and new product performance: Towards the understanding of the mechanisms and key contingencies', *Research Policy*, 47 (2), 498–510.

Murray, N., Lynch, P. & Foley, A. (2016) 'Unlocking the magic in successful tourism destination marketing: the role of sensing capability'. *Journal of Marketing Management*, 32 (9-10), 877-899.

Nieves, J. & Diaz-Meneses, G. (2018), 'Knowledge sources and innovation in the hotel industry: Empirical analysis on Gran Canaria Island, a mature mass-tourism destination', *International Journal of Contemporary Hospitality Management*, 30 (6), 2537-2561.

Nieves, J. & Haller, S. (2014) 'Building dynamic capabilities through knowledge resources', *Tourism Management*, 40, 224-232.

Omerzel, D. G. (2016) 'A systematic review of research on innovation in hospitality and tourism', *International Journal of Contemporary Hospitality Management*, 28 (3), 516-558.

Otengei, S. O., Bakunda, G., Ngoma, M., Ntayi, J. M. & Munene, J. C. (2017) 'Internationalization of African-ethnic restaurants: A qualitative enquiry using the dynamic capabilities perspective', *Tourism Management Perspectives*, 21, 85-99.

Pace, L. A. (2016) 'How do tourism firms innovate for sustainable energy consumption? A capabilities perspective on the adoption of energy efficiency in tourism accommodation establishments', *Journal of Cleaner Production*, 111, 409-420.

Parjanen, S., Hennala, L. & Konsti-Laakso, S. (2012) 'Brokerage functions in a virtual idea generation platform: Possibilities for collective creativity?', *Innovation*, 14 (3), 363-374.

Paswan, A. K., D'Souza, D. & Rajamma, R. K. (2014) 'Value co-creation through knowledge exchange in franchising', *Journal of Services Marketing*, 28 (2), 116–125.

Peronard, J. P. & Brix, J. (2019) 'Organizing for inter-organizational learning in service networks', *Learning Organization*, 26 (3), 276–288.

Pinto, H., Cruz, A. R. & Combe, C. (2015) 'Cooperation and the emergence of maritime clusters in the Atlantic: Analysis and implications of innovation and human capital for blue growth', *Marine Policy*, 57, 167-177.

Pittz, T. G., Intindola, M. L., Adler, T., Rogers, S. & Gard, C. (2019) 'Collaborating Smartly: The Role of Open Strategy in Absorptive Capacity', *Journal of Small Business Management*, 57 (4), 1595–1615.

Prior, D. D., Keränen, J. & Koskela, S. (2018) Sensemaking, sensegiving and absorptive capacity in complex procurements, *Journal of Business Research*, 88, 79-90.

Pucci, T., Runfola, A., Guercini, S. & Zanni, L. (2018), 'The role of actors in interactions between "innovation ecosystems": Drivers and implications', *IMP Journal,* 12 (2), 333-345.

Randhawa, K., Wilden, R. & Gudergan, S. (2018) 'Open Service Innovation: The Role of Intermediary Capabilities', *Journal of Product Innovation Management,* 35 (5), 808-838.

Randhawa, K., Wilden, R. & Hohberger, J. (2016) 'A bibliometric review of open innovation: Setting a research agenda', *Journal of Product Innovation Management,* 33 (6), 750-772.

Rodríguez, A. Barón, N. & Martinez, J. (2020) 'Validity of dynamic capabilities in the operation based on new sustainability narratives on nature tourism SMEs and Clusters', *Sustainability,* 12, 1004-1022.

Sarker, S., Sarker, S., Sahaym, A. & Bjørn-Andersen, N. (2012) 'Exploring value co-creation in relationships between an ERP vendor and its partners: A revelatory case study', *M I S Quarterly,* 36 (1), 317-338.

Shaw, G. & Williams, A. (2009) 'Knowledge transfer and management in tourism organisations: An emerging research agenda', *Tourism Management,* 30 (3), 325-335.

Su, C.-Y., Lin, B.-W. & Chen, C.-J. (2016) 'Knowledge co-creation across national boundaries: trends and firms' strategies', *Knowledge Management Research & Practice,* 14 (4), 457–469.

Seuring, S. & Gold, S. (2012) 'Conducting content-analysis based literature reviews in supply chain management', *Supply Chain Management,* 17 (5), 544-555.

Taheri, M. & van Geenhuizen, M. (2016) 'Teams' boundary-spanning capacity at university: Performance of technology projects in commercialization', *Technological Forecasting and Social Change,* 111, 31–43.

Thomas, R. (2012) 'Business elites, universities and knowledge transfer in tourism', *Tourism Management,* 33 (3), 553-561.

Thomas, R. & Ormerod, N. (2017) 'The (almost) imperceptible impact of tourism research on policy and practice', *Tourism Management,* 62, 379-389.

Thomas, R. & Wood, E. (2014) 'Innovation in tourism: Re-conceptualising and measuring the absorptive capacity of the hotel sector', *Tourism Management,* 45, 39-48.

Thomas, R. & Wood, E. (2015) 'The absorptive capacity of tourism organisations', *Annals of Tourism Research,* 54, 84-99.

Tussyadiah, L. & Zach, F. (2014). Capacity for co-creation among destination marketing organizations. In S. McCabe (ed.) *The Routledge Handbook of Tourism Marketing* (pp. 425-435), London: Routledge.

Van Geenhuizen, M. & Nijkamp, P. (2012) 'Knowledge virtualization and local connectedness among young globalized high-tech companies', *Technological Forecasting and Social Change*, 79 (7), 1179-1191.

West, J. & Bogers, M. (2014) 'Leveraging external sources of innovation: A review of research on open innovation', *Journal of Product Innovation Management*, 31 (4), 814–831.

Wilden, R., Gudergan, S., Akaka, M. A., Averdung, A. & Teichert, T. (2019) 'The role of cocreation and dynamic capabilities in service provision and performance: A configurational study', *Industrial Marketing Management*, 78, 43–57.

Wilke, E. P., Costa, B. K., Freire, O. B. D. & Ferreira, M. P. (2019) 'Interorganizational cooperation in tourist destination: Building performance in the hotel industry', *Tourism Management*, 72, 340-351.

Zahra, S. A. & George, G. (2002) 'Absorptive capacity: A review, reconceptualization, and extension', *The Academy of Management Review*, 27 (2), 185-203.

Zavattaro, S. M., Daspit, J. J. & Adams, F. G. (2015) 'Assessing managerial methods for evaluating place brand equity: A qualitative investigation', *Tourism Management*, 47, 11-21.

3 Social interaction in co-creating the tourist experience: An exploratory study of Chinese visitors to Japan

Xing Han, Carolus L.C. Praet and Liyong Wang

The objectives of this chapter are to:

- ☐ Explore the nature of social interaction as a necessary condition for value co-creation in tourism.
- ☐ Address the lack of studies on the role of social interaction in co-creating the tourist experience among tourists from non-Western cultures.
- ☐ Investigate how frequently Chinese tourists have social interactions with three groups of social actors: service providers, residents, and other tourists.
- ☐ Clarify how Chinese tourists evaluate the importance of social interactions with each of the three groups of social actors.
- ☐ Discuss the theoretical and managerial implications of our findings, the limitations of our study, and suggest avenues for further research.

Keywords: co-creation; social interaction; tourism experience

This study was supported by JSPS Kakenhi Grant Number 18K11864. Xing Han acknowledges the Fuji Xerox Kobayashi Fund Research Grant Program for Foreign Doctoral Candidates in Japan, and Toshimi Otsuka Scholarship Foundation (Japan) for the financial assistance during this study.

Introduction

Tourists are increasingly looking for experiences that allow them to actively participate (Campos et al., 2015), and the role of the tourist as a co-creator, or even the sole creator, of the tourism experience is becoming widely recognized among tourism scholars and practitioners (Binkhorst & Dekker, 2009; Rihova et al., 2013; Campos et al., 2015). Some even regard active value co-creation as the most important aspect of the tourism experience (e.g.Boswijk, Thijssen & Peelen, 2007; Binkhorst & Dekker, 2009).

Destinations may be viewed as a space in which tourists co-create their own memorable experiences through social and other interactions (Morgan & Xu, 2009). Following this conceptualization, social interaction among participants is an essential condition for value co-creation. Categories of social actors participating in tourism value co-creation include service providers (e.g., Salvado, 2011; Minkiewicz, Evans & Bridson, 2014), the local community (e.g., Azevedo, 2009; Richards, 2010), and other tourists (e.g., Rihova et al., 2013, 2015; Reichenberger, 2017).

Surprisingly perhaps, previous tourism studies have tended to take social interactions among actors in tourism for granted, and have often treated them as inherently positive. Hence, we feel that an investigation of these basic assumptions of positive social interactions and how they are perceived by tourists is warranted. Furthermore, tourists' perceptions and attitudes towards different types of on-site social actors may lead to different forms of value co-creation. We therefore need to study from the perspective of the tourist how different types of social actors participate in the tourism value co-creation process, and what is the relative impact of each type of social actor on this process.

Literature review

The co-creation of tourist experiences

Since the end of the 20th century, the economies of developed nations have started shifting towards what Pine and Gilmore (1999) call the 'experience economy'. The experience economy designates a shift away from 'modern' economies where consumers merely consume goods and services toward 'post-modern' economies where consumers look for more unique, personalized, hedonic, emotional, aesthetic, or cultural/educational experiences, in the production of which they actively engage. Academic studies are increasingly focusing on the consumer's role as a co-creator of experiences.

Some even argue that it is the very experience of the co-creation itself that consumers desire and value (Prahalad & Ramaswamy, 2003, 2004; Binkhorst & Dekker, 2009). Prahalad and Ramaswamy (2003) call for a new perspective where customers actively co-create unique value for themselves through personalized interaction. In a similar vein, Vargo and Lusch (2004, 2008) suggest that value is always co-created through active and dynamic firm-consumer interaction and integration of resources of the supply and demand sides.

While the phenomena of tourism, travel and leisure have always incorporated experiential aspects to a higher extent than other types of services and goods, the focus on the more experiential aspects of travel among tourism and leisure scholars is relatively new (Scott, Laws & Boksberger, 2009).

The co-creative aspect of a typical tourism experience means that it usually involves frequent social interactions (Li & Petrick, 2008; Maunier & Camelis, 2013). As a result, some scholars highlight the creation of meaningful and unique value for individual tourists through personal interaction and call for a deeper understanding of tourist social interaction (Rihova et al., 2013; Campos et al., 2015). Social actors who participate in tourist social interaction (and thus co-create tourist experiences) can be categorized into three types: service providers (e.g., Salvado, 2011; Minkiewicz, Evans & Bridson, 2014), the local community (e.g., Azevedo, 2009; Richards, 2010), and other tourists (e.g., Rihova et al., 2013, 2015; Reichenberger, 2017). To better understand the tourist experience, it is necessary to know what kind of role these three groups of social actors play in the co-creation of tourist experiences. In the following sections, we will review the extant literature on the role of each of these three groups of social actors in the tourist experience co-creation process.

Social interaction with service providers

A fresh look at the role of service providers in co-creating unique and memorable tourist experiences places tourists at the center of their own experiences. Arnould and Price (1993) argue that the interaction and relationship between the tourist and service personnel directly influence the tourist's emotional reactions to extraordinary experiences. In this perspective, the front-line employee becomes an important operant resource for co-creating customer experiences, and eventually contributes to improving the organization's competitive advantage (Lusch, Vargo & O'Brien, 2007).

Several studies have explored the role various types of tourism service providers play in the tourism experience across different tourism contexts.

For example, Ap and Wong (2001) regard the tour guide as one of the key front-line staff members who may potentially transform a tour into an experience for customers. Arnould and Price's study (1993) on river rafting suggests that the tour guide plays a subtle yet important role in delivering an extraordinary experience. Mossberg (1995) also reports that tourist satisfaction with the tourism experience is largely influenced by the performance of the tour guide. In addition to the tour guide, other types of hospitality front-line personnel also play an essential role in creating the tourist experience. Lashley (2008) states that hospitality is essentially a relationship based on hosts and guests. Similarly, Hemmington (2007) stresses that the main distinctive characteristic of hospitality lies in the host-guest relationship.

Social interaction with residents

Residents are an indispensable part of the tourism destination and thus play an important role in the social aspects of tourist experiences. The local community may function as an appealing tourism attraction for the tourist. Morgan and Xu (2009) suggest that tourist interaction with the local culture and people contributes to a unique and memorable tourism experience. Kim's study (2010) confirms this by reporting that the local culture of a destination, such as the friendliness of locals, is one of the determinants that make a travel experience memorable. Brown's study (2005) on volunteer tourism reveals that the desire to immerse oneself physically and emotionally in the local culture and community is a strong motivation for tourists.

The notion that the hospitality of residents is of vital importance for the tourism industry and the tourist is widely accepted (e.g., Bimonte & Punzo, 2016; Lin, Chen & Filieri, 2017). The support and goodwill of the local population lead to the success of tourism development, and are equally vital to creating positive and memorable experiences for tourists. In contrast, negative or hostile attitudes towards tourists may destroy a destination's tourism value and are likely to discourage tourists' willingness to interact with the local community. Huang and Hsu (2005) report that mainland Chinese tourists to Hong Kong sensed the negative and self-superiority attitudes of the residents towards them, and suggest that such negative feelings may discourage these tourists' intention to revisit. Also, Li et al. (2011) point out that the perceived discrimination of Chinese people among Australians was one of the constraining factors influencing their motivation to travel to Australia. In contrast, other studies report that the perceived friendliness and hospitality of Japanese towards Chinese people reduce anxiety among

Chinese visitors when traveling in Japan (Ji, Li & Hsu, 2016; Lin, Chen & Filieri, 2017).

Appreciation of the important role of residents in the tourism experience has led researchers to investigate resident perceptions of the impacts of tourism and attitudes towards co-creating tourist experiences (Gursoy, Jurowski & Uysal, 2002; Bimonte & Punzo, 2016; Lin, Chen & Filieri, 2017). This topic is widely explored from the perspective of social exchange theory, which suggests that people are likely to engage in an exchange if they believe that they can gain benefits without incurring unacceptable costs (Gursoy, Jurowski & Uysal, 2002; Rasoolimanesh et al., 2015). According to this perspective, the decision to participate in social exchange with visitors depends on residents' belief that the benefits of doing so will outweigh the costs.

Bimonte and Punzo (2016) find that the tourist-host interaction may influence host attitudes, perceptions, and lifestyles as well as tourist perceptions and satisfaction. Fan et al. (2016) also report that tourist-host interaction decreases the perceived cultural distance with the host culture among tourists. For this reason, they recommend destination management to involve residents into the local tourism industry by encouraging them to actively interact with tourists (Fan et al., 2016).

Social interaction with other tourists

Tourism experiences typically take place in the presence of and/or in collaboration with other tourists. As a result, the social interactions and shared experiences with other tourists may constitute an important part of the tourism experience. Moreover, the increasing number of global tourists is likely to intensify the impact of tourist-tourist interaction on the tourism experience.

Grove and Fisk's (1997) study on tourist experiences at theme parks found that the behavior of other tourists influenced tourists' overall evaluation of the experience. The study concluded that 'protocol incidents', i.e., when tourists must share time and space and must follow expected rules of conduct, may influence tourist experiences positively or negatively (Grove & Fisk, 1997). Alternatively, 'sociable incidents' may establish temporary friendships among tourists (Arnould & Price, 1993; Grove & Fisk, 1997). Huang and Hsu (2010) studied interaction between tourists on cruise vacations and confirmed the positive effect of tourist-tourist interaction on cruise experience and vacation satisfaction. Moreover, several studies point out that sometimes not only direct interaction, but even the mere presence of other tourists may impact the tourist experience (Grove & Fisk, 1997; Yagi, 2001; Praet et al., 2015).

An increasing number of studies acknowledge the important role of other tourists and address tourist-tourist interaction from the perspective of co-creation (Prahalad & Ramaswamy, 2004; Scott, Laws & Boksberger, 2009; Reichenberger, 2017). The experience co-created between tourists entails their active involvement and thus results in higher levels of satisfaction, word-of-mouth, perceived value, and loyalty (Campos et al., 2015; Reichenberger, 2017). At the same time, tourist behaviors to create value for themselves may intentionally or inadvertently come at the expense of the experience/perceived value of other customers and thus may lead to a diminishment of the value created for these customers (e.g., Grove & Fisk, 1997; McColl-Kennedy & Tombs, 2011).

Research purpose

A review of the tourism literature suggests that the few studies on tourist social interaction have tended to focus on Western tourists and that they generally pay no attention to tourists from non-Western cultures, including Chinese tourists, an important and ever-growing source market for outbound tourists in global tourism (Li et al., 2013; WTO, 2018).

The purpose of this study is to address the nature of social interaction as a necessary condition for value co-creation in tourism. As a first step, we focus on how tourists perceive these social interactions. More specifically, we study how Chinese tourists perceive social interactions with service providers, residents, and other tourists. Finally, we explore the perceived importance of social interaction with each of these three groups of social actors in co-creating the tourism experience.

Methodology

We collected data through in-depth interviews. We developed a preliminary list of interview questions based on a review of the relevant academic literature in addition to an analysis of Chinese tourist-generated online content regarding personal travel experiences. We then made additional modifications to the interview questions based on five pilot interviews.

We conducted 29 interviews with 42 interviewees during June to July in 2018 in situ to enable more effective elicitation of tourist emotions and meaningful memories (Campos et al., 2015). Nineteen interviews took place in a tourist information center in the city of Sapporo in Northern Japan and ten at a popular Shinto shrine in Tokyo.

We digitally recorded the interviews with the consent of the interviewees. The lead author and one of the co-authors shared the task of interviewing visitors from mainland China in Chinese. The lead author then transcribed all the interview recordings into verbatim text excluding the paralanguage. The co-author who had also conducted the interviews then randomly verified 20% of the transcripts to confirm their accuracy and consistency.

The lead author was in charge of coding the transcripts. Before starting the formal coding, she first read the transcripts repeatedly to gain a thorough understanding of the data (Miles & Huberman, 1994). As the second step of the analysis she adopted the manual coding technique following Strauss and Corbin's (1990) approach, which includes open coding, axial coding, and selective coding. The back-and-forth coding process resulted in a mature coding framework, and she then transferred the data into the NVivo 12 software package, which she used to generate the final coding framework and to re-code the original transcripts accordingly.

Results

Analysis of the interview transcripts identified 162 reported cases of interaction with three types of social actors: service providers, residents and other tourists. Our identification of three types of social actors in tourism value co-creation confirms findings of previous studies on this topic.

The reported cases include what we have labeled 'direct' and 'indirect' social interaction. The naming and basic conceptualization of 'direct' versus 'indirect' interaction follows prior studies on tourist-tourist interaction in tourism (e.g., Huang & Hsu, 2009; Huang & Hsu, 2010; Kim & Lee, 2012; Yang, 2015). While previous studies have not elaborated on how 'direct' and 'indirect' social interaction differ, the current study conceptualizes and expands these two types of social interaction to include all social actors participating in the creation of the tourism experience, i.e., service personnel, residents, and other tourists, as follows. Direct social interaction occurs when social actors notice and respond to each other's presence by means of verbal or non-verbal 'outward' or 'observable' communication. Outward communication takes the form of showing that one has noticed or recognized the other person by making a gesture or by other ways of greeting. In contrast, indirect social interaction occurs when social actors notice the presence of other social actors without any direct or observable communication or response taking place between the tourist and the other social actors. Indirect, or 'inward' interaction may thus occur only inside the mind of the social actors and does not require any outward forms of communication to occur. It is noteworthy that about 20% of the interactions reported (see Table 3.1) are indirect or 'inward' interactions, which implies that other people may influence the tourist experience without any direct interaction taking place.

Our data also suggest that tourist interactions with different groups of social actors manifest themselves in different ways, and that tourists hold different attitudes towards the interactions depending on the type of social actors involved. Table 3.1 summarizes the relative number of the reported interactions with different types of social actors, and interviewee attitudes towards these interactions, from positive to negative. We will describe our findings in more detail in the next sections.

Table 3.1: Frequency and evaluation of interactions with other social actors

Type of interaction	Evaluation	Service providers		Residents		Tourists		Subtotal	
		n	%	n	%	n	%	n	%
Direct interaction	Positive	21	57	29	83	10	17	60	
	Neutral	4	11	5	14	47	81	56	
	Negative	12	32	1	3	1	2	14	
	Subtotal	37	100	35	100	58	100	130	80
Indirect interaction	Positive	3	60	0	0	3	12	6	
	Neutral	0	0	2	100	10	40	12	
	Negative	2	40	0	0	12	48	14	
	Subtotal	5	100	2	100	25	100	32	20
Total		42		37		83		162	100

Source: Authors' elaboration.

Social interactions with service providers

Most direct interactions with service providers reported by the interviewees occurred during routinely performed and high-protocol service encounters. A relatively small number of reported cases of direct tourist-service provider interaction took place when the tourist asked for help from the service provider, e.g., when asking the service staff for directions. Interviewees also reported indirect interactions with service providers. The indirect interactions refer to two types of situations: 1) when interviewees reported they were amazed by the excellence of the service provider's performance, and 2) when interviewees observed how service providers were dealing with other tourists.

Overall, tourist-service provider interaction largely takes place in relatively manualized, scripted situations. Nevertheless, not only did most of our interviewees (n=15) appear to attach a relatively high importance to interaction with service providers but they also reported having a relatively strong emotional involvement with it. The strength of the emotional involvement can be inferred from the finding that positive (57%) and negative (32%) direct interactions accounted for a combined 89% of all service provider-related comments. One possible reason for this is that, for the typical tourist, interaction with service providers is unavoidable. In contrast, it is easier for tourists to avoid interactions with other tourists. The following is a typical quote:

> *"I value the interaction with service providers most, because you always have to interact with them. Regarding the tourists, if you don't like them you can just walk away and it's totally OK."* (Female, FIT, 20s).

Other interviewees attached importance to interactions with service providers because they believed that the interactions are closely connected to their personal interest.

> *"You are just an outsider for other tourists when interacting with them. Whereas it is your own tangible benefit that is connected to interactions with the service provider."* (Female, FIT, 30s).

Regarding the incidents when tourists asked for help from service providers, interviewees viewed service staff not only in their role of professionals who provide services of a limited range and within the specific context of a service encounter, but also as a reliable source of general help beyond the range of the service that their job descriptions require them to provide.

> *"If I did not prepare for the trip well enough, I would choose to ask the service staff for help. ... The residents here may not be able to help us, unless they are also interested in traveling like us."* (Female, FIT, 30s).

One interviewee called the service providers the 'window' of the destination.

> *"The service provider is just like the window [of the destination]. Our first choice is always the service provider (when we need help). They are our first impression of the destination."* (Female, FIT, 20s).

Several of our interviewees used the word 'window' as a metaphor for the role of the service provider. For most interviewees, service providers are the ones they most frequently interact with, either passively or actively. Moreover, interviewees placed their trust and personal interest in the interactions with the service staff. Hence, how tourists evaluate the destination is closely related to the professionalism and overall demeanor of service staff members. The response of one interviewee is a good illustration:

> *"The most social interactions we have are with service providers and that's why we care about them. Because we need to interact with them constantly, one negative interaction out of ten is significant enough to change our impression of the service [of the destination]. Regarding residents here, we have too few interactions to care about that. Also, we cannot request local people to cater to our needs."* (Male, FIT, 30s).

Social interactions with residents

Tourist-resident interaction was the least common type of interaction mentioned by the interviewees (n=34). Tourist interactions with residents were typically a result of tourists requesting help from residents, or of residents proactively offering help to tourists.

Consistent with previous studies suggesting that tourist-host interaction is an important part of the tourism experience (e.g., Brown, 2005; Fan et al., 2016), most of the interviewees expressed positive attitudes toward direct interactions with residents (83% of all reported direct interactions with residents). Residents, in addition to the personnel providing services to tourists, constitute the authentic part of the social environment of the destination and they consequently have a big influence on the tourism experience. Greenblat and Gagnon (1983) suggest that social interaction with host people helps to alleviate the tourist's anxiety from being a 'temporary stranger' in an unfamiliar environment. Interviewees in this study also expressed similar opinions.

> *"It is the local people who have a bigger influence after all, including both the common resident and the service provider. After setting foot on the land of the destination, if the service you receive and the people you encounter treat you welcomingly, it will make the surroundings less unfamiliar and help you better fit in the local atmosphere. You will like the place more."* (Female, group tour, 30s).

Specifically, interviewees value the genuine human touch through interactions with local people. Interviewees compared their interactions with residents to the interactions they had with service providers and tended to express a preference for interactions with residents because they viewed the interactions with service providers as mechanical and obligatory.

> *"The service staff serve one wave of tourists after another and their service is kind of emotionless. However, with the local folks you can have a much sincerer interaction."* (Female, group tour, 20s).

Tourists are attracted to social interactions with residents mainly due to the desire for getting closer to the authentic everyday lives of local people. Urry and Larsen (2011) suggest that tourists particularly show fascination with the "real lives" of others (p. 10). The following are two representative quotes which support this assertion:

> *"We probably would choose to patronize places with more local people. Because we believe that the locals would know better, for example, about [good] restaurants. Also, we want to experience the local life more."* (Male, FIT, 30s).

> *"[To have] contact with the local people is the reason we travel. Tourism is all about [experiencing] the local environmental conditions and customs, which means that apart from the scenery, [experiencing] the customs of local people is equally important. The most important is the human context."* (Female, FIT, 40s).

While the search for authentic experiences might be a universal motive for traveling, only some of our interviewees expressed the desire to interact with residents directly. For most interviewees, the language barrier and the lack of opportunity prevented them from directly interacting with local people.

> *"Traveling is more about experiencing the local environmental conditions and customs. However, it's not that easy to open up and communicate frankly with local people because of the language barrier. Therefore, when it comes to the local life, it's more about the environmental conditions, whereas the experience of local customs is very much limited. Because I feel that there are quite significant cultural differences, wherever you go. It's always difficult to sit down and have a conversation, unless you are really proficient in the local language."* (Female, FIT, 40s)

While they expressed a fascination with authentic social interactions with local people, interviewees at the same time were aware of the difficulties and barriers that prevented them from having this kind of interaction. One interviewee (Female, FIT, 20s) stated that having a genuine direct interaction with local people is a 'bonus' while traveling. Several of our interviewees said they often compromised on their desire to have genuine interaction with residents and instead would quietly observe the local way of life without having to interact verbally with the host community. This behavior, i.e., the tourist observing local people from a distance, is what Urry (1990) calls the 'tourist gaze'.

Social interactions with other tourists

Among all social interactions reported by the interviewees in this study, interactions with other tourists were those that they most frequently mentioned (n=83). It suggests that other tourists are the most salient type of social actor that influences the tourism experience. At the same time, interviewees viewed the majority of direct interactions with other tourists as having little impact on their experience; while 81% of the comments describing interactions with other tourists could be classified as having a neutral impact, only 14% of interactions with residents and 11% of interactions with service providers could be categorized as neutral.

Whereas most interviewees thus appeared to be somewhat indifferent to direct interactions with other tourists, they did report a considerable number (n=25) of indirect tourist-tourist interactions, among which 48% could be classified as negative (vs. 12% positive and 40% neutral). The finding of indirect interaction with other tourists as an influencing factor confirms Yagi and Pearce's (2007) suggestion that even the mere presence of other tourists could potentially have an impact on the tourism experience.

This study also confirms that tourists do not always consider the role of other tourists as a positive factor in the co-creation of tourism experiences. Interviewees held mixed views of the roles other tourists play in tourism experiences. Some tourists held a positive attitude towards other tourists and valued the tourist-tourist interaction as they were accompanying each other temporarily in the destination.

> *"I went on a one-day tour in Okinawa .... We [I and other tourists] had lunch and went to the aquarium together [in a group] .... We talked to each other occasionally, but [all of these conversations] were very brief."* (Female, FIT, 30s)

In contrast, for some tourists the appealing factor when visiting a tourism attraction is the absence of other tourists. Some interviewees tended to view other tourists as a competitor for tourism resources and tourism services. This suggests that Chinese outbound tourists are more inclined to create their own experiences without the involvement of other tourists. In this regard, the role of other tourists in the tourist experience is not one of co-creating, but rather one of diminishing the experience.

> *"You just need to avoid the high season. When you go to the tax refund or somewhere else, it will be full of tourists and you need to wait for an hour to get it done."* (Female, FIT, 20s)

The interview data also reveal the existence of a special segment of 'other tourists', i.e., other Chinese tourists. Chinese outbound tourists appear to embrace mixed feelings towards their compatriots when traveling. On the one hand, when many tourists from China visit the same destination this may diminish the exotic atmosphere, whereas on the other hand meeting compatriots and communicating with them in Chinese also reconciles the anxiety of coping with an unfamiliar environment. The followings are two typical illustrations of Chinese tourists' mixed views towards other Chinese tourists.

> *"I feel that Chinese are everywhere. Whenever I am shopping or doing something else, I always have the urge to go over and say 'hi' to the Chinese tourists I see. I just have an amiable feeling towards them."* (Female, group tour, 40s)

> *"I come to Japan to relax but it turns out that this place has been occupied by Chinese.... If I were surrounded by Japanese tourists, and I was the only Chinese tourist, I would feel much better. Because I come to Japan to escape my familiar environment and to relax..."* (Female, FIT, 20s)

Conclusion

This study explored Chinese tourist perceptions of the role of social interaction in the co-creation of tourism experiences with three types of social actors: service providers, residents and other tourists. In doing so, this study contributes to the literature on co-creation in tourism, not only by exploring this under-researched topic, but also by clarifying the relative importance and nature of tourist social interactions with the three types of social actors, from a non-Western perspective.

First, our data show that tourist interactions with service providers appear to have an important impact on the tourism experience for Chinese tourists to Japan. Interviewees appeared to have a stronger emotional involvement toward interactions with service providers, since both positive and negative interactions account for a considerable portion of reported incidents.

Second, our interviewees mostly looked forward to social interactions with residents. Compared to the ritualized and manualized nature of the interaction with service providers, Chinese tourists perceive interaction with local people as more sincere, authentic, and desirable. This appears to confirm Urry and Larsen's (2011) suggestion that tourists show particular fascination with the 'real lives' of others. Nevertheless, even though most of the interviewees looked for authentic experiences in Japan, only a few expressed the desire to directly interact with residents. The main barriers to direct interaction with residents were the perceived lack of opportunity to come into contact with them and the difficulty of communication in a foreign language. Tourism destinations could eliminate these barriers by organizing cultural exchange activities and by using translation technology.

Third, and contrary to previous studies which suggest that the desire to come into contact with other tourists is one of the main motivations for traveling (Pearce, 2005a; de Rojas & Camarero, 2008), this study found that Chinese tourists try to avoid rather than to meet other tourists: avoidance of other tourists not only happens at the time of selecting a travel destination but also upon arrival at the destination. The reason for this may be that tourists tend to view other tourists as competitors (Pearce, 2005b) for tourism services, or as inhibiting the exotic or authentic atmosphere of the destination. These findings suggest that Chinese tourists are not only more inclined to create their own experiences without the involvement of other tourists, but even perceive the role of other tourists in co-creating the experience as largely negative, i.e., as diminishing the experience. Despite their desire to avoid other tourists, our interviewees at the same time expressed an inability to do so, since most of the reported interactions (both direct and indirect) were those with other tourists (n=84). Moreover, these tourist-tourist interactions even outnumbered the reported interactions with service providers (n=42) and residents (n=34). This finding raises the issue of customer compatibility. Previous studies (e.g., Lazarevski & Dolnicar, 2008; Harris, 2013) have addressed customer compatibility from a service organization point of view, whereas the findings of this study highlight the customer perspective. For tourism destinations to better manage customer compatibility, a deeper understanding of tourist perceptions and attitudes towards other tourists would appear crucial. Based on insights regarding perceived compatibility of inter- and intra-national customer segments, destination and hospitality managers can put in place segmentation and targeting strategies that create a better balance in customer nationality. In addition, accommodation managers could introduce several interface design strategies that foster positive social interaction or avoid negative interaction among tourists from either the same or other nationalities.

Limitations and avenues for future research

While this study provides several new insights, it also has a number of limitations. One limitation arises from the definition of 'indirect interaction'. This study tentatively uses indirect interaction to refer to the 'inward' interaction may thus occur only inside the mind of the social actors and does not require any outward forms of communication to occur. This definition follows previous studies on tourist-tourist interaction, which have used the term 'indirect interaction' as opposed to 'direct interaction', to refer to the influence of the presence of other tourists, customer density, crowding and public behaviors of other tourists on the tourism experience (e.g., Huang & Hsu, 2009, 2010; Kim & Lee, 2012; Yang, 2015). We feel that more discussion is needed on whether the term 'indirect interaction' is appropriate to describe a situation when no mutual, physical inter-action, but only a unilateral, mental reaction occurs.

Another limitation of the study is that it was based on a relatively small number of interviews, and also that its findings may only be valid for Chinese tourists visiting Japan. Further qualitative and quantitative studies involving tourists of Chinese and other nationalities, and in other destinations, are needed.

In contrast to findings of previous studies among Western tourists which show that social interaction with other tourists is an important motivation for traveling, our study shows that this appears to be less important for Chinese tourists. It is necessary to further investigate what are the factors that influence Chinese tourist preferences to socialize or to avoid other tourists. Future research should also be conducted on tourists from various cultural backgrounds to explore how and to what extent cultural background influences attitudes, inclinations and behaviors regarding social interaction in creating experiences with other tourists, or whether this is more a function of the individual tourist's demographic characteristics, personality, motiviation for traveling, and travel style.

References

Ap, J. & Wong, K. K. F. (2001) 'Case study on tour guiding: Professionalism, issues and problems', *Tourism Management*, 22 (5), 551-563.

Arnould, E. & Price, L. (1993) 'River magic: Extraordinary experience and the extended service encounter', *Journal of Consumer Research*, 20 (1), 24-45.

Azevedo, A. (2009) 'Designing unique and memorable experiences: Co-creation and the "surprise" factor', paper presented at the III Congresso Internacional de Turismo de Leiria e Oeste - 2009, Leiria, Portugal.

Bimonte, S. & Punzo, L. F. (2016) 'Tourist development and host-guest interaction: An economic exchange theory', *Annals of Tourism Research*, 58, 128-139.

Binkhorst, E. & Dekker, T. D. (2009) 'Agenda for co-creation tourism experience research', *Journal of Hospitality and Leisure Marketing*, 18 (2-3), 311-327.

Boswijk, A., Thijssen, T. & Peelen, E. (2007) *The Experience Economy, A New Perspective*, Amsterdam: Pearson Education Benelux.

Brown, S. (2005) 'Travelling with a purpose: Understanding the motives and benefits of volunteer vacationers', *Current Issues in Tourism*, 8 (6), 479-496.

Campos, A. C., Mendes, J., do Valle, P. O. & Scott, N. (2015) 'Co-creation of tourist experiences: A literature review', *Current Issues in Tourism*, 21 (4), 369-400.

de Rojas, C. & Camarero, C. (2008) 'Visitors' experience, mood and satisfaction in a heritage context: Evidence from an interpretation center', *Tourism Management*, 29 (3), 525-537.

Fan, D. X. F., Qiu, H., Jenkins, C. L. & Lin, P. M. C. (2016) 'Does social contact lessen perceived cultural distance? Evidence from tourist-host social contact', *Annals of Tourism Research*, 61, 223-225.

Greenblat, C. S. & Gagnon, J. H. (1983) 'Temporary strangers: Travel and tourism from a sociological perspective', *Sociological Perspectives*, 26 (1), 89-110.

Grove, S. J. & Fisk, R. P. (1997) 'The impact of other customers on service experiences: A critical incident examination of "getting along"', *Journal of Retailing*, 73 (1), 63-85.

Gursoy, D., Jurowski, C. & Uysal, M. (2002) 'Resident attitudes: A structural modeling approach', *Annals of Tourism Research*, 29 (1), 79-105.

Harris, A. (2013) 'Like me or like me? The role of compatibility and similarity in backpacker networks', paper presented at the 19th Asia Pacific Tourism Association Annual Conference, 1-4 July, Bangkok, Thailand.

Hemmington, N. (2007) 'From service to experience: Understanding and defining the hospitality business', *The Service Industries Journal*, 27 (6), 747-755.

Huang, J. & Hsu, C. H. (2009) 'Interaction among fellow cruise passengers: Diverse experiences and impacts', *Journal of Travel & Tourism Marketing*, 26 (5-6), 547-567.

Huang, J. & Hsu, C. H. (2010) 'The impact of customer-to-customer interaction on cruise experience and vacation satisfaction', *Journal of Travel Research*, 49 (1), 79-92.

Huang, S. S. & Hsu, C. H. (2005) 'Mainland Chinese residents' perceptions and motivations of visiting Hong Kong: Evidence from focus group interviews', *Asia Pacific Journal of Tourism Research*, 10 (2), 191-205.

Ji, M., Li, M. & Hsu, C. H. C. (2016) 'Emotional encounters of Chinese tourists to Japan', *Journal of Travel & Tourism Marketing*, 33 (5), 645–657.

Kim, J. H. (2010) 'Determining the factors affecting the memorable nature of travel experiences', *Journal of Travel and Tourism Marketing*, 27 (8), 780-796.

Kim, N. & Lee, M. (2012) 'Other customers in a service encounter: Examining the effect in a restaurant setting', *Journal of Services Marketing*, 26 (1), 27-40.

Lashley, C. (2008) 'Marketing hospitality and tourism experiences', in H. Oh and A. Pizam (eds.), *Handbook of Hospitality Marketing Management*, Oxford: Butterworth-Heinemann, pp. 3-31.

Lazarevski, K. & Dolnicar, S. (2008) 'Tourist segment compatibility', paper presented at the conference Australian and New Zealand Marketing Academy (ANZMAC), 1-3 December, Sydney.

Li, X. (Robert), Meng, F., Uysal, M. & Mihalik, B. (2013) 'Understanding China's long-haul outbound travel market: An overlapped segmentation approach', *Journal of Business Research*, 66 (6), 786-793.

Li, X. & Petrick, J. F. (2008) 'Tourism marketing in an era of paradigm shift', *Journal of Travel Research*, 46 (3), 235-244.

Li, M., Zhang, H., Mao, I., & Deng, C. (2011) 'Segmenting Chinese outbound tourists by perceived constraints', *Journal of Travel and Tourism Marketing*, 28 (6), 629-643.

Lin, Z., Chen, Y. & Filieri, R. (2017) 'Resident-tourist value co-creation: The role of residents' perceived tourism impacts and life satisfaction', *Tourism Management*, 61, 436-442.

Lusch, R. F., Vargo, S. L. & O'Brien, M. (2007) 'Competing through service: Insights from service-dominant logic', *Journal of Retailing*, 83 (1), 5-18.

Maunier, C. & Camelis, C. (2013) 'Toward an identification of elements contributing to satisfaction with the tourism experience', *Journal of Vacation Marketing*, 19 (1), 19-39.

McColl-Kennedy, J. R. & Tombs, A. (2011) 'When customer value co-creation diminishes value for other customers deliberately or inadvertently', paper presented at the conference Naples Forum on Service: Service Dominant Logic, Network & System Theory and Service Science, 15-17 June, Capri.

Miles, M. B. & Huberman, A. M. (1994) *Qualitative Data Analysis: An Expanded Sourcebook*, 2nd edn, Thousand Oaks: SAGE Publications.

Minkiewicz, J., Evans, J. & Bridson, K. (2014) 'How do consumers co-create their experiences? An exploration in the heritage sector', *Journal of Marketing Management*, 30 (1-2), 30-59.

Morgan, M. & Xu, F. (2009) 'Student travel experiences: Memories and dreams', *Journal of Hospitality and Leisure Marketing*, 18 (2-3), 216-236.

Mossberg, L. (1995) 'Tour leaders and their importance in charter tours', *Tourism Management*, 16 (6), 437-445.

Pearce, P. L. (2005a) 'The role of relationships in the tourist experience', in W. F. Theobald (ed), *Global Tourism, 3rd edn*, Amsterdam: Butterworth-Heinemann, pp. 103-122.

Pearce, P. L. (2005b) *Tourist Behavior: Themes and Conceptual Schemes*, Clevedon: Channel View Publications.

Pine, B. & Gilmore, J. H. (1999) *The Experience Economy*, Boston: Harvard Business School Press.

Praet, C. L. C., Gotoh, H., Miyazaki, Y., Lee, J. & Wang, L. (2015) 'Familiar and total strangers: An exploratory study of in-group and out-group perceptions among international ski resort visitors', in A. Ono, M. Kikumori & R. Takeuchi (eds.), *Emerging Trends in Asian Markets, Proceedings of the 2015 International Conference of Asian Marketing Associations*, Tokyo: Japan Society for Marketing and Distribution, pp. 120-135.

Prahalad, C. K. & Ramaswamy, V. (2003) 'The new frontier of experience innovation', *MIT Sloan Management Review*, 44 (4), 12-18.

Prahalad, C. K. & Ramaswamy, V. (2004) *The Future of Competition: Co-creating Unique Value with Customers*, Boston: Harvard Business School Press.

Rasoolimanesh, S. M., Jaafar, M., Kock, N. & Ramayah, T. (2015) 'A revised framework of social exchange theory to investigate the factors influencing residents' perceptions', *Tourism Management Perspectives*, 16, 335-345.

Reichenberger, I. (2017) 'C2C value co-creation through social interactions in tourism', *International Journal of Tourism Research*, 19 (6), 629-638.

Richards, G. (2010) 'Tourism development trajectories - From culture to creativity?' *Tourism & Management Studies*, 6, 9-15.

Rihova, I., Buhalis, D., Moital, M. & Gouthro, M. B. (2013) 'Social layers of customer-to-customer value co-creation', *Journal of Service Management*, 24 (5), 553-566.

Rihova, I., Buhalis, D., Moital, M. & Gouthro, M. B. (2015) 'Conceptualising customer-to-customer value co-creation in tourism', *International Journal of Tourism Research*, 17 (4), 356-363.

Salvado, J. (2011) 'Ecosystem model: Building travel agencies' business resilience in Portugal', *European Journal of Tourism, Hospitality and Recreation*, 2 (1), 95-116.

Scott, N., Laws, E. & Boksberger, P. (2009) 'The marketing of hospitality and leisure experiences', *Journal of Hospitality and Leisure Marketing*, 18 (2-3), 99-110.

Strauss, A. L. & Corbin, J. M. (1990) 'Grounded theory research: Procedures, canons, and evaluative criteria', *Qualitative Sociology*, 13 (1), 3-21.

Urry, J. (1990) *The Tourist Gaze Leisure and Travel in Contemporary Societies*, London: Sage.

Urry, J. & Larsen, J. (2011) *The Tourist Gaze 3.0*, London: Sage.

Vargo, S. L. & Lusch, R. F. (2004) 'Evolving to a new dominant logic for marketing', *Journal of Marketing*, 68 (1), 1-17.

Vargo, S. L. & Lusch, R. F. (2008) 'Service-dominant logic: Continuing the evolution', *Journal of the Academy of Marketing Science*, 36 (1), 1-10.

WTO (World Tourism Organization) (2018) 'Asia and the Pacific newsletter (2018 - issue 46)', Retrieved July 3, 2018 from http://cf.cdn.unwto.org/sites/all/files/pdf/180608unwto46jeoyongryang.pdf.

Yagi, C. (2001) 'How tourists see other tourists? Analysis of online travelogues', *Journal of Tourism Studies*, 12 (2), 22-31.

Yagi, C. & Pearce, P. L. (2007) 'The influence of appearance and the number of people viewed on tourist preferences for seeing other tourists', *Journal of Sustainable Tourism*, 15 (1), 28-43.

Yang, F. X. (2015) 'Tourist co-created destination image', *Journal of Travel & Tourism Marketing*, 33 (4), 425-439.

4 Emotional interactions in festivals: How do consumers build a collective emotional experience?

Nico Didry and Jean-Luc Giannelloni

The objectives of this chapter are to:

- ☐ Deepen the analysis of the collective emotional dimension at play during a consumer experience.
- ☐ Show that the nature of emotional interactions and their intensity is a central component of the experience.
- ☐ Identify consumption strategies and logics implemented by festival-goers to develop this collective dimension of the experience.
- ☐ Propose a new framework to analyze emotional interactions in a collective recreational context, in order to consider all types of emotion transfers, not reduced to the visitors' companions.
- ☐ Propose the notion of emotional group in order to take into consideration unexpected and ephemeral encounters that are looked for during this leisure experience.

Keywords: collective emotional dynamics; consumer behavior; festival; tourist experience

Introduction

Music festivals are factors of attractiveness for territories. As such they are part of their tourism strategies (Getz, 1991). In France, 84% of the 2018 music festivals took place during the touristic summer season. They sometimes even become a tourist product in itself like Tomorrowland Winter in Alpe d'Huez, a ski resort in the French Alps. During seven days, the ski resort is only accessible for the festival-goers. In 2019, Alpe d'Huez was fully filled with 23,000 tourists from 131 different countries who booked their holidays to enjoy skiing and concerts during that special event, and 36,000 people were on the waiting list.

Provoking a spatio-temporal rupture with everyday life (Chaney, 2011), significant in leisure or tourist practices, festivals allow experiencing a real re-enchantment of the world and everyday life. According to the post-modern approach, the phenomenon of society around festivals, illustrated by the growth in festival demographics (in the number of participants but also in the number of events) (Négrier et al., 2013), can be considered in the global context of a return to festive alchemy and the cult of pleasure, with a powerful return to affect and emotion. This festival craze is significant for the "triumph of the collective will to live over the individual" (Maffesoli, 2012: 115). However, this collective dimension of emotions has received limited attention in marketing (Didry & Giannelloni, 2019). In addition, although accompaniment has often been analyzed in consumer behavior (Debenedetti, 2003), few studies consider the collective context in which consumers are immersed in their experience. If a festival experience is lived in a collective way, which behaviors do festival consumers develop to engage in emotional interactions with others? The challenge here is to bring a new reading of the experience of collective consumption through emotional transfers to fill a gap in the marketing literature. More specifically, it is a question of assessing how the need for emotional interactions will influence the festival-goer's behavior.

Theoretical framework and contribution

The literature shows that the festival experience must be seen from a social and emotional perspective. A festival is to be approached through the notions of link (Cova, 1995) and community, insofar as it responds, through the value of the link, to the need of 'postmodern' individuals to satisfy their desire for community (Badot & Cova, 2009). Participation in a festival is then seen as a tribal phenomenon (Cova & Cova, 2002) in which festivals constitute "cultural transhumance, or festive nomadism" (Maffesoli, 1988). In these tribes, even ephemeral ones, what will prevail is to be connected to the other, "to develop a common feeling and, together, make it the heart of these gatherings" (Ferrand, 2009: 30).

Collective effervescence (Durkheim, 1912), through emotional communion, can, from a postmodern approach, be considered as a central, even as an essential component of the festival. Consumption experience depends on the nature and intensity of the emotions people experience (Addis & Holbrook, 2001). Festivals are synonymous of sharing emotions and bonds, and festival-goers come with their companions (Collin-Lachaud, 2010). According to Rimé (2007), emotional experiences are systematically shared with others. Sharing emotions is an opportunity sought by individuals even more so that living emotions for themselves (Rimé, 2005). Any emotion expressed in the presence of others can trigger emotional interactions (Barsade, 2002). Pulh et al. (2005) highlighted the value of social links in the experience of social connection at three levels, the value of social interaction, social practice, and social communion. We propose here to deepen this analysis from an emotional point of view to fill this gap in the marketing literature.

That is why we will focus on the processes of emotion transfer that can explain festival-goers' behaviors. There are two types of emotional diffusion processes within groups (Didry & Giannelloni, 2019):

1) the social sharing of emotions, which is a voluntary evoking of emotions in a socially shared form (Rimé, 2005); and
2) the phenomenon of emotional contagion, which is a tendency to automatically synchronize facial and vocal expressions, postures and movements with those of another person and, consequently, to converge emotionally (Hatfield et al., 2009).

These two concepts relating to the diffusion of emotions are placed within the overall collective emotional dynamics' framework (Didry & Giannelloni, 2019). Those emotional transfers have a direct effect on the ambiance of the audience, also called 'emotional atmosphere', which is a short-term emotional state resulting from the average emotional responses of members of a group to a common event (De Riviera, 1992).

From a conceptual point of view, it emerges that the social and cultural dimensions are inherent in the concept of emotion. These emotions are a real communication tool for social regulation and cultural identification. The processes of exchange and diffusion of emotions (sharing of emotions, emotional contagion), whether conscious or automatic, are correlated to social interactions, as well as systematically reinforcing social bonds (Luminet, 2008; Rimé, 2005; Tcherkassof, 2008). Thus, in the same way that Tarde (1901) proposed to move from a psychology of crowds to a sociology of crowds, it seems appropriate to move on from a psychology of emotions to a sociology of emotions, which makes it possible to approach collective consumption contexts with a more appropriate theoretical perspective.

Even if there is no truly universal definition of the 'emotion' construct (Derbaix & Poncin, 2005) that would make sense for the different currents, whether in marketing or in the disciplines that have studied this phenomenon for more than a century, we will retain the broad and consensual definition of Tcherkassof (2008: 15) which defines emotion as "an episode that emerges automatically, imposing its precedence over any other current activity, an episode made up of a set of interrelated and synchronized responses, of a neuro-psychological, expressive behavioral, subjective-phenomenal, cognitive-attentional and motivational type". We, therefore, understand emotion as a multi-component phenomenon, a true social regulator, with reference to current multi-component theories (Scherer, 1984; Frijda, 2007; Coan, 2010; Tcherkassof & Frijda, 2014) proposing a perceptive model of emotion. This perceptive model, which emphasizes the expression of emotions, allows socio-cultural data to be taken into account. Moreover, the notion of context, particularly social context, is essential for understanding emotions. As there is a clear consensus in the literature that emotions are a true social regulator, emotions should therefore not be viewed as individual phenomena, as it is usually the case (Smith, Seger & Mackie, 2007).

Method and materials

As the objective of our research is to understand how festival consumers live their experience in terms of emotional transfers, we have adopted an interpretive approach whose purpose is to give meaning to the observed behaviors, through qualitative methods. This interpretive approach is privileged by the Consumer Culture Theory (CCT) trend (Ozcaglar-Toulouse & Cova, 2010), which focuses on the individual's experience. The interpretive perspective advocates the study of the consumer by 'immersion' (Bergadaà, 1990), which we have implemented through an ethnographic approach in the sense of Desjeux (1990) and Badot et al. (2009). This approach provides an anthropological view of consumption, i.e. a cultural analysis of the observed phenomena (Desjeux, 1990). The objective, through the use of this method, is to understand the consumption experience from the inside (Robert-Demontrond et al., 2013), which is particularly suited to our subject; in our case, it is not just a matter of going towards 'real people' (Askegaard & Linnet, 2011), but of living with them and like them.

The data were collected during three different music festivals, in order to establish a significant comparative approach to multi-site ethnography (Hannerz, 2003). The three music festivals chosen for the data collection were Hadra Trance Festival 2013 and 2014 (international psytrance music festival in France gathering more than 16,000 festival-goers during a four-day non-stop music performance), Musilac Festival 2016 and 2017 (pop-rock festival in Aix-les-Bains, France with 50,000 people) and Tomorrowland Winter Festival 2019 (first edition of the winter version of the famous international electro-music festival, that took place in Alpe d'Huez, ski resort in the French Alps, attracting 23,000 festival-goers, 78% of whom came from outside France).

The ethnographic immersion is characterized by the plurality of sources but also of data collection methods (Kozinets, 2009). Photos, videos, note-taking, logbook, semi-directive, or informal interviews were performed during each event (Table 4.1).

Table 4.1: Data collection tools used in this research

Hadra Trance Festival 2013	logbook, semi-structured interviews (7), informal interviews (7), introspection, videos, photos, document collect.
Hadra Trance Festival 2014	logbook, informal interviews (12), longitudinal observation of one festival-goer, introspection, videos, photos, document collect
Musilac Festival 2016	logbook, videos, photos, introspection
Musilac Festival 2017	logbook, videos, photos, introspection, semi-structured interviews (2)
Tomorrowland Winter Festival 2019	logbook, semi-structured interviews (5) informal interviews (11), introspection, videos, photos

Source: Authors' elaboration.

Data were also collected afterward on social media (through the Facebook pages of the events) to collect festival-goers' verbatim according to the netnographic approach (Kozinets, 2009). A specific analysis of the traditional post-event festival organizer 'thank you' publication's comments was conducted. The netnography of post-festival comments has been considered as an asynchronous observation in the sense of Dion and Sitz

(2013), as the traces left by consumers. Comments, but also posted videos, were collected and analyzed.

Textual data were coded according to an open thematic manual process (Corbin & Strauss, 1990; DeWalt & DeWalt, 2011). We used a semiologic analysis for visual data (Beaud & Weber, 2017). Then, the data processing was carried out using a triangulation process involving all the data. Gradually, an analysis grid was refined. It integrated the components of the festival experience in terms of sharing emotions, but also the dimensions of the experience more broadly as well as cultural, social and community aspects.

Results and discussion

The quality and nature of emotional interactions as an experience barometer

Our results show above all that the phenomena of emotional contagion and the sharing of emotions are omnipresent within the festival. Everyone is subject to the emotions of others, confirming Barsade's (2002) results in the working place context. For some, the nature of social interactions is even the most crucial point regarding the experience: "Thank you to all those festival-goers I met and who smiled to me, because it was in these smiles that I found the most wonderful resource of energy and happiness!!!!!!! " (Publication of a festival-goer of the Hadra Festival). The 'People of Tomorrow' also show a strong interest in maintaining this goodwill and respect between people that will condition the quality of the emotional atmosphere (De Riviera, 1992). The notion of exchange is essential in this experience: "when everyone talks with everyone" (Mary). It is "the right encounters" (Mélina) that make the consumer say that he is having a good time. The comparative netnographic analysis of comments to post-event Facebook 'thank you' publications from festivals confirms this. Indeed, the results of our netnographic study show that the sources of satisfaction of the experience are the emotional and social interactions with the others festival-goers (Hadra Trance Festival and Tomorrowland Festival), or the artist performance when the spatial organization of the festival and the culture of the community do not facilitate the setup of social interactions (Musilac Festival). For the participants, the human factor is systematically put forward to characterize the satisfaction of the experience and the success of the event. On the other hand, the human aspect is mainly responsible for significant inconveniences during the event: "respect is apparently not a notion shared by everyone!" (Aurore, in an angry tone, claiming in a Facebook publication after Musilac Festival).

Strategies of festival-goers looking for emotional interactions

Festival-goers develop strategies to live a fun and shared experience and create the conditions for transferring emotions to other festival-goers. We have identified three strategies, which we're going to present on a passive-proactive continuum:

1. Removing barriers to emotion transfer

This strategy consists of setting up structural conditions to facilitate emotional exchange between two or more people. We observed two forms:

1) Physical barriers removal to allow facial expression of emotions (Tcherkassof, 2008). Our observations during festive gatherings on Tomorrowland Winter show that when festival-goers entered a festive zone (outdoor dance floor) they removed almost systematically the ski equipment hiding their face (e.g. goggles, neckwarmer or sunglasses) even if the weather conditions were not good. Removing the ski-goggles or neck warmer helps to create the conditions for the exchange, to show that the festival-goer is willing or waiting to exchange.
2) Mental barriers removal to disinhibit behavior and thus break down barriers between festival-goers. The psychotropic drugs (alcohol, drugs) taken by the festival-goers during those collective and festive moments have this function. The common point of the effects sought by those who use drugs is the removal of inhibitions and/or the desire to communicate. Thus, taking drugs and alcohol, because of the decrease of shyness or social modesty, facilitates this contact and the intimacy of the relationship: "it's easier to meet people, we know it, it makes it easier" (Rémi during the Hadra Trance Festival). This allows people both to reach out more easily to others (contaminator), but also to develop the attitude of emotional openness allowing to be contaminated more easily.

Our observations show that this way of facilitating emotional openness by removing barriers to the transfer of emotion is most of the time an unconscious strategy or at least an automatism from the festival-goers.

2. Dress with disguise or accessories

The goal is to encourage social interactions. "It's contagious, we attract a lot of people with our disguises" (Acajou / HTF). During the day, colorful, atypical or exuberant clothing is legion among festival-goers and encourages social interaction. They are a pretext to initiate a discussion or simply to exchange a smile. "It's to get noticed, attract attention and to discuss with people" (Nicolas, HTF organizer). During day and night festivals, like the Hadra Trance Festival, they change at night for fluorescent colors or phosphorescent accessories to remain visible. The creation of social interactions between festival-goers who do not know each other, through disguises, is a first step in the process of establishing emotional interactions. The disguise serves as a pretext to establish an exchange between people. This two-step process is significant to this dressing-strategy.

Costumes and accessories are not only linked to a notion of identity (Masset & Decrop, 2017), but also to a desire to create social interactions in order to share emotions with new people.

Four types of disguises were identified to define four festival-goer profiles regarding the use of costumes to create social and emotional interactions. All of them encourage emotional exchanges, both outside and inside the group.

The playful

In this category we find those who disguise themselves in a playful logic with original, offbeat and differentiating costumes or those who divert the use of accessories, inflatable objects for swimming pools for example, or as in Figure 4.1 with the little horse which systematically provokes social interactions and exchanges.

Figure 4.1: Social interactions 'creation due to accessories (one minute between the two photos) / Hadra Trance Festival 2014. Source: Nico Didry.

The patriot

Those who wear their national flag to show where they come from. "With my flag, I've made a lot of friends, people come up to me and ask me if I'm from South Africa, and why there's an Australian flag on the bottom of my South African flag" (Bradley, TML). It's also part of the Tomorrowland codes and culture and thus specific to the Tomorrowland Festival. The organizers enhance the fact that there are 113 different nations for the winter edition and more than 200 for the summer edition, to show its worldwide attractivity. This culture or behavior to wear one's country or regional flag is starting to spread in other festivals (Figure 4.2.).

Figure 4.2: Flag and nations costumes during Tomorrowland Winter Festival 2019. Source: Nico Didry.

The community based

In this profile, festival-goers are showing their belonging to the festival community or the specific music community of the festival, wearing t-shirts, caps, make-up, or accessories like the festival's flag or the music style community. The festival access bracelets collection proudly displayed on one's wrist is a way to show this belonging and involvement but also a way to initiate social interactions. For the Hadra Trance Festival or Tomorrowland, the festival-goer shows that he knows the codes of the festival, specifically the notions of respect and sharing and that she/he is open to sharing emotions with others. For the more classical festivals (e.g. mainstream festival without a strong link with a community or without sharing values), showing one's belonging to the community highlights similarities and common points, and allows the creation of a synergy between the members of the community, who recognize each other, which helps to initiate relations (Figure 4.3).

Figure 4.3: Festival or community costumes during Tomorrowland Winter Festival 2019 and Harda Trance Festival 2014. Source: Nico Didry.

The group-based

Those who choose to wear the same costumes or accessories as the other members of the group. Our observations highlight that this behavior fosters emotional and social exchanges within the group (intra-group). This is the main motivation of their members: "We want to live something even stronger together" (Sam, TML). But it also creates strong social interactions with other groups (inter-group) when two disguised groups are meeting. Even if it is less conducive to interpersonal exchanges if the group remains centered on itself and is not open to the outside world, it helps to create a festive atmosphere, free of social modesty, facilitating emotional interactions (Figure 4.4.).

Figure 4.4: Group costumes during Tomorrowland Winter Festival 2019. Source: Nico Didry.

3. Implementation of actions or behaviors to reach out to others and create relationships

This is the more proactive strategy, in which the festival-goers reach out to one another. They use accessories like sprays, make-up, or smartphones to do selfies. These will allow the contaminator to go directly to others rather than waiting for others to come to him. The emblematic example is the spray or vaporizer, very present in the daytime and when it is hot (Figure 4.5.). Community members use it to create the first eye contact and to be able to exchange smiles. "It's a way to meet people. At the festival, you always have people who are dressed up, or with spray, make-up. It creates exchanges: bringing something to the other, and having something in exchange, a look, a thank you " (Emilie, HTF).

Figure 4.5: Spray or vaporizer to initiate emotional exchanges during Hadra Trance Festival 2013. Source: Nico Didry.

Finally, we also observe even more direct attitudes such as the hug, "this is a trance thing, there is no barrier, we hug, take care of you" (Emilie, HTF). This exchange of emotion involving direct physical contact shows a strong closeness between the members of the community. Hugs are not reserved for people who already belong to the circle of friends or relatives but are used between people who do not know each other. "This morning, a guy was really happy and he was kissing everybody" (Alex, HTF). This behavior was observed very frequently during the Hadra Festival. This closeness and this opening of the festival-goers, via the hug in particular, were not observed in other places. Only the "free hug" behavior (people carrying a sign with "free hugs" written on it) proposed by festival-goers was observed in other festivals (Musilac, TML). But the approach was not proactive as at the Hadra Trance Festival. This confirms the importance of the cultural aspect of the context via values and social codes (Figure 4.6.).

Figure 4.6: Hugs between festival-goers who don't know each other during Hadra Trance Festival 2013. Source: Nico Didry.

All these behaviors shape the festival experience. Even if all the festival-goers do not have any proactive and conscious strategy, they are all affected by the pro-active behavior of the festival-goers. The proportion of festival-goers who are pro-active in creating emotional exchanges varies greatly from one festival to another and depends in particular on the cultural codes of the community linked to the festival.

Three levels of emotional interactions between festival goers

Our research shows that the experience is strongly shaped by its human dimension and particularly its emotional factor. However, this has three levels:

1) The restricted circle – "the people I am with, the people around me" (Acajou). At this level, the diffusion of emotions, through the social sharing of emotions or through interpersonal emotional contagion, between people in the intimate sphere of the spectator are omnipresent and continuous throughout the experience. These emotional exchanges are interpersonal.
2) The ephemeral and random encounters – "the people we will meet" (Jeremy). This ranges from an exchange of smiles with a person crossed on the festival to a more intense sharing of emotions in the case of the abovementioned hugs, or when two groups are meeting for example.
3) The extended circle – "the people who dance" (Alex), i.e. the active audience in general. The attractiveness of the crowd observed at festivals is significant at this level.

Groups move from one place to another according to the number of people present in that place, and not only according to the artistic content. The broader human environment is considered by festival-goers and is part of the strategies linked to emotional exchanges: "It's very different if there's no crowd; it's like a thing between friends, more private. It's more intimate, (...) but you have no one to share with" (Aymeric). Festival-goers seek to experience emotional interactions with the crowd as an emotional unit (Le Bon, 1895). Emotional contagion is not only interpersonal but should also be extended to a wider environment. We suggest integrating this 'mass' contagion (Didry & Giannelloni, 2014) systematically in the collective emotional dynamics' studies.

Thus, it shows that the three levels of social link value proposed by Pulh et al. (2005) (i.e. the value of social interaction, social practice, and social communion) can be analyzed on an emotional dimension.

Finally, our study allows us to deepen and refine the categorizations of social interactions during the consumption experience. Roederer (2012) has identified three categories of people with whom the consumer interacts during the consumer experience: family members (friends, family, colleagues...), strangers (crowds, other store customers, etc.) and contact staff (guides, controllers, etc.). Our results show that this categorization does not work in the context of the festival experience, mainly because it does not allow for the dissociation of crowds and unexpected encounters.

Broadening the notion of accompaniment, towards an emotional group

During the Hadra Festival, many festival-goers come alone because they know that they will have social and emotional interactions with lots of festival-goers during the festival. There is less need to be accompanied as many festival-goers become ephemeral or longer-term companions, like Jean-Baptiste who says "no worry, you walk 20 metres and you're meeting someone". The netnographic data analysis of the post-event comments on the Facebook page of the Hadra Trance Festival shows that many festival-goers went alone to the festival and that many encounters lasted during the whole festival. This calls into question the fact that the lack of companion is a major barrier to participation, as Searle and Jackson (1985) have suggested. This also leads us to suggest that accompaniment may not be necessary. Debenedetti (2003: 44) defines accompaniment of the visitor to a leisure place as the "presence or absence of one or more relatives at his or her side (spouse, friends, family member) during his or her trip to that place". However, our study suggests that this notion of accompaniment should be broadened in order to approach this notion with a dynamic perspective of emotion, in the sense that, for example, companions can change during the experience.

Drawing on the concept of a group of friends (Decrop et al., 2007), we suggest an extension of the group units to approach the study of entertainment leisure activities. Although Debenedetti (2003) had already proposed to consider the group rather than the individual, arguing that in leisure activities the unit of analysis is not the individual but the intimate social group to which he belongs, our study shows that in the specific context of festivals, this is not enough. The group of friends is, according to Jehn and Shah (1997), a group with close and interpersonal ties and positive and friendly pre-existing relationships between members. We propose to add as an analytical unit, the groups resulting from unexpected encounters at the place of experience that we will qualify as an 'emotional group'. By emotional group, we mean a group that is formed transiently to live the experience together and share its emotions through direct interpersonal interactions. We also suggest integrating as an analytical unit the group in the broad sense present at the festival, i.e. the public in the sense of Tarde (1901), with whom the festival-goer is linked by a sense of belonging and identification. We used the framework of Decrop et al. (2007) to characterize the different types of groups, to develop a typology of groups to be considered as a unit of analysis in the context of festivals and shows.

Conclusion

Our study of emotional interactions during music festivals highlights the need to take into account the wider human environment in order to analyze this consumer experience and the behavior of spectators or festival visitors. Beyond the fact that the nature and quality of the social and emotional interactions lived during the experience are determining factors in consumer satisfaction, it emerges that these emotional exchanges are sought after by festival-goers. This search for social contact and emotional exchange helps to explain their behavior. Different strategies are used, ranging from passive to proactive, which also highlights different profiles of festival-goers in terms of their search for emotional interaction. Some prefer to be contaminated and then share their emotions in return, others prefer to be contaminators and then receive emotions. Our results emphasize that this festival experience is not only a complete collective experience but also a collaborative experience. The emotional atmosphere depends directly on the social and emotional interactions developed by the festival-goers. Music festival managers and touristic product managers should consider this, and develop strategies to increase the collective emotional dimension of the experience, by implementing codes, values or actions that facilitate social and emotional interactions.

We proposed an analytical grid based on the observation of social and emotional interactions with three types of subjects: (1) groups of friends, accompanying persons with whom they had social links prior to the event, and with whom the festival-goer will share most of the festival experience; (2) emotional groups, persons met in an ephemeral and random way during the festival. We have observed that the strategies implemented by festival-goers (disguises, etc.) to create encounters that are oriented towards this type of person; (3) the crowd, an aggregate of people perceived as a mental unit, also expressing their emotions.

The phenomenon of emotional contagion, therefore, occurs at two levels, an interpersonal level, and a mass level.

We have also shown that an analysis of the collective emotional dimension of an event makes it possible to explain the accompanying phenomena. Therefore, accompaniment attributes are directly related to the possibility of sharing emotions and experiencing emotional interactions with others during the experience. It opens a new field of research to determine those specific characteristics.

The strategies developed by the festival-goers, and their behaviors regarding interpersonal relationships are varied and can be different from one festival to another as they are linked to the cultural codes related to the festival. Therefore, the nature and intensity of emotional exchanges also depend directly on the socio-cultural context, and on the codes and values put forward by the community present at the festival.

References

Addis, M. & Holbrook, M.B. (2001) 'On the conceptual link between mass customisation and experiential consumption: An explosion of subjectivity', *Journal of Consumer Behaviour*, 1 (1), 50-66.

Askegaard, S. & Linnet, J.T. (2011) 'Vers une épistémologie de la théorie de la culture du consommateur (CCT): phénoménologie, structure et contexte du contexte', *Perspectives culturelles de la consommation*, 1 (1), 15-42.

Badot, O. & Cova, B. (2009) *Néo-marketing [reloaded]*, Editions EMS.

Badot, O., Carrier, C., Cova, B., Desjeux, D. & Filser, M. (2009) 'L'ethnomarketing: un élargissement de la recherche en comportement du consommateur à l'ethnologie', *Recherche et Applications en Marketing*, 24 (1), 94–111.

Barsade, S. G. (2002) 'The ripple effect: Emotional contagion and its influence on group behavior', *Administrative Science Quarterly*, 47 (4), 644–675.

Bergadaà, M. (1990) 'The role of time in the action of the consumer', *Journal of Consumer Research*, 17 (3), 289-302.

Beaud, S. & Weber, F. (2017) 'Guide de l'enquête de terrain: produire et analyser des données ethnographiques', *La Découverte*.

Chaney, D. (2011) 'Conquête ou fidélisation, la perception des stratégies relationnelles d'un festival de musiques actuelles par ses organisateurs et son public', *Décisions Marketing*, 63, 49-59.

Coan, J.A. (2010) 'Emergent ghosts of the emotion machine', *Emotion Review*, 2, 274-285.

Collin-Lachaud, I. (2010) 'Eurockéennes, Francofolies, Vieilles Charrues ou Main Square Festival: Le rituel communautaire comme source de fidélisation', *Décisions Marketing*, 60, 43-53.

Corbin, J. & Strauss, A. (1990) 'Grounded theory research: Procedures, canons, and evaluative criteria', *Qualitative Sociology*, 13 (1), 3-21.

Cova, B. (1995) 'Au-delà du marché: Quand le lien importe plus que le bien', Paris, L'Harmattan.

Cova, V. & Cova, B. (2002) 'Les particules expérientielles de la quête d'authenticité du consommateur', *Décisions Marketing*, 28, 33-42.

Debenedetti, S. (2003) 'L'expérience de visite des lieux de loisirs: Le rôle central des compagnons', *Recherche et Applications en Marketing*, 18 (4), 43-58.

Decrop, A., Pecheux, C. & Bauvin G. (2007) 'La prise de décision dans les groupes d'amis: Une étude exploratoire', *Recherche et Applications en Marketing*, 22 (2), 1-21.

Derbaix C. & Poncin I. (2005) 'La mesure des réactions affectives en marketing: évaluation des principaux outils', *Recherche et Applications En Marketing*, 20 (2), 55–75.

De Rivera, J.H. (1992) 'Emotional climate: Social structure and emotional dynamics' in K.T. Strongman (ed.), *International Review of Studies on Emotion*, New York: John Wiley & Sons, pp. 197–218.

Desjeux, D. (1990) 'L'ethnomarketing, un nouveau souffle pour les études de comportement des consommateurs?', *Sciences Humaines*, 1, 38-39.

DeWalt, K.M. & DeWalt, B.R. (2011) *Participant Observation: A Guide for Fieldworkers*, Rowman Altamira.

Dion D. & Sitz L. (2013), 'Enrichir la compréhension des comportements des consommateurs : pistes opérationnelles et enjeux organisationnels', *Décisions Marketing*, 71, 45-58.

Durkheim, E. (1912/1984) *Les formes élémentaires de la vie religieuse*, Paris: Presses Universitaires.

Didry, N. & Giannelloni, J-L. (2014) 'Les émotions partagées lors d'une expérience collective de consommation: Approche socioculturelle du Hadra Trance Festival', *Actes des 13èmes journées Normandes de la Recherche sur la Consommation*, Rouen.

Didry, N. & Giannelloni, J-L. (2019) 'Collective emotional dynamics: Perspectives for marketing', *Recherche et Applications en Marketing*, 34 (4), 99-124.

De Rivera, J.H. (1992) 'Emotional climate: Social structure and emotional dynamics' in K.T. Strongman (ed.), *International Review of Studies on Emotion 2*, New York: JohnWiley&Sons, pp. 197–218.

Ferrand, L. (2009) 'Comprendre les effervescences musicales. L'exemple des concerts de rock', *Sociétés*, 104, 27-37.

Frijda, N. (2007) *The Laws of Emotion*, New York: Lawrence Erlbaum Associates Publishers.

Getz, D. (1991) *Festivals, Special Events and Tourism*, New York: Van Nostrand Reinhold.

Hannerz, U. (2003) 'Being there... and there... and there!: Reflections on multi-site ethnography', *Ethnography*, 4 (2), 201-216.

Hatfield E., Rapson R.L. & Le Y.L. (2009) 'Emotional contagion and empathy', in J. Decety and J. Ickes (coord), *The Social Neuroscience of Empathy*, p19-30, Boston, MIT Press.

Jehn, K.A. & Shah, P.P. (1997) 'Interpersonal relationships and task performance: An examination of mediation processes in friendship and acquaintance groups', *Journal of Personality and Social Psychology*, 72 (4), 775.

Kozinets, R.V. (2009) *Netnography: doing ethnographic research online*, London: Sage Publications.

Le Bon, G. ([1895]/1963) *La psychologie des foules*, Paris: PUF.

Luminet, O. (2008) *Psychologie des émotions*, Bruxelles: Ed de Boeck.

Maffesoli, M. (1988) *Le temps des tribus: Déclin de l'individualisme dans les sociétés postmodernes*, Paris: Méridiens.

Maffesoli, M. (2012) *Homo eroticus. Des communions émotionnelles*, Paris: CNRS Éditions.

Masset, J. & Decrop, A. (2017) 'Tomorrowland festival: A heterotopia of deviation (30min)'. *Advances in Consumer Research*, 45.

Negrier, E., Guerin, M. & Bonet, L. (coord.) (2013) *Festivals de musiques, un monde en mutation*, Paris: Michel de Maule.

Ozcaglar-Toulouse, N. & Cova, B. (2010) 'Une histoire de la CCT française, parcours et concepts clefs', *Recherche et Applications en Marketing*, 25 (2), 69-91.

Pulh M., Bourgeon D. & P. Bouchet (2005) 'Spectacles vivants, logiques de consommation et construction d'expériences : le paradoxe d'une offre à la fois unique et plurielle', *Décisions Marketing*, 37, 57-66.

Rimé, B. (2005), *Le partage social des émotions*, Paris: PUF.

Rimé, B. (2007) 'The social sharing of emotion as an interface between individual and collective processes in the construction of emotional climates', *Journal of Social Issues*, 63 (2), 307-322.

Robert-Demontrond, P., Joyeau, A., Beaudouin, V., Bellion A. & Sugier L. (2013) 'De l'Odyssée à l'Illiade: Stratagèmes et compétences pratiques en ethnomarketing', *Recherche et Applications en Marketing*, 28 (4), 103-127.

Roederer, C. (2012) *Marketing et consommation expérientiels*, France: EMS management et société.

Searle, M. & Jackson, E. (1985) 'Socioeconomic variations in perceived barriers to recreation participation among would-be participants', *Leisure Sciences*, 7 (2), 227-249.

Scherer, K.R. (1984) 'Les émotions, fonctions et composantes', *Cahiers de la psychologie cognitive*, 4 (1), 9-39.

Smith, E.R., Seger, C., & Mackie, D. (2007) 'Can emotions be truly group level? Evidence regarding four conceptual criteria', *Journal of Personality and Social Psychology*, 93 (3), 431-446.

Tarde, G. (1901) *L'opinion et la foule*, Paris: Alcan.

Tcherkassof, A. (2008) *Les émotions et leurs expressions*, Grenoble: PUG.

Tcherkassof, A. & Frijda, N. (2014) 'Les émotions, une conception relationnelle', *L'année psychologique*, 114 (03), 501-535.

Part 2
Platforms and the Collaborative Economy

5 Collaborative economy in the tourism industry: The new deal for consumers in the European Union

Silvana Canales Gutiérrez

The objectives of this chapter are to:

- ☐ Identify the main elements of the collaborative economy and their impact on tourism in the European Union.
- ☐ Point out the benefits of the sharing economy for tourists in the European Union.
- ☐ Indicate the nature of the guest-tourist as a consumer and holder of special rights in the European Union.
- ☐ Determine consumer rights in the European Union regarding international jurisdiction and applicable law.
- ☐ Analyze whether the clauses of international jurisdiction and applicable law of the service contracts established by the platforms are adapted to consumer rights in the European Union.

Keywords: collaborative economy, European Union, consumer protection law, consumer rights.

Introduction

Europe is the most touristic continent in the world, receiving more than 50% of all international tourists (Santolli, 2017) according to the World Tourism Organization. People from all over the world want to go to the most famous tourist attractions in Europe and what once seemed a distant dream to international tourists due to the high prices of hotels and air tickets, is now possible thanks to the competitive prices of international airlines such as Ryanair, Vueling and EasyJet (O'Connell & Williams, 2005) and the alternative to traditional accommodation providers: collaborative economy platforms such as Airbnb, HomeAway or Wimdu. This short research paper will be focused on this type of platform, which provide mainly hosting services, and the legal aspects of their terms and conditions of service.

The collaborative economy in the tourism industry is a growing business model, which allows consumers around the world to rent a spare room, an entire house or an apartment, for a short period of time, at a lower price than the accommodation offered by the traditional service providers such as hotels. However, this phenomenon was not born as a trending idea or an alternative way of getting an extra income, but of the pure necessity of generating cash in a period when the economy was stagnating, and the owners of properties needed to be creative with the available resources. The collaborative economy is characterized by generating economic benefit (Botsman & Rogers, 2010) from assets that would otherwise be given little or no use by their owners or holders. However, the concept of 'resources' covers much more than just assets, since resources can refer to spaces, skills and any kind of goods, which, if not made available to the collaborative economy, would be largely unused.

This legal analysis was carried out in 2019. The clauses that were taken from the terms and conditions of service of the Airbnb, Toprural, Homelidays and Escapade platforms to carry out this analysis have changed since that date, as well as the their corporate composition; their closing or merged to another company, and any other external factor after December 30, 2019 was not taken into account to perform this analysis. This is underpinned by the fact that the reality of the platforms develops in a more agile and changing way that the study on their legality, however, this analysis sets out the situation of the four platforms chosen up to a specific date by setting a precedent for their compliance or non-compliance with European Union consumer protection regulations.

It is necessary to clarify that the platforms are not lessors, they do not own the properties to be rented; they are merely intermediaries between the owners of the properties and the future guests (Fainmesser, 2014). Their main service is to act as intermediaries for the provision of short-term lodging, but they can also provide underlying services that can be connected to the lodging or can be totally independent (European Commission, 2016). These underlying services range from making reservations in different types of establishments to booking adventure experiences, sports activities or activities related to the local culture. The platforms are set up in accordance with a sales strategy that suggests tourists must have a complete experience in the place where they are staying, and they offer tourists options that allow them to get to know the place they are visiting quickly and on a budget.

From the tourist's position, the growth of the accommodation on offer was a refreshing alternative to hotels; together with the newest trend in property matters: using and enjoying the property, without even thinking about acquiring it. Therefore, from a tourist decision-making point of view, selecting specific accommodation depends on two basic conditions, the location and the price, which are directly related to the level of comfort you seek for your stay. The first condition, the location, is key to determining if the potential guest chooses a property offered by the collaborative economy in the tourism industry. The proximity of tourist areas, the availability of efficient means of transport, the physical conditions of the roads and the local aid available to tourists are essential factors when choosing a tourist destination and later planning a holiday. Regarding the second condition, the price, it is reasonable to point out that the primary fixed costs such as public services and taxes are borne by the owner, because in order to be the holder of the property, he is responsible for at least the maintenance of the dwelling.

Thus, the owner of the property obtains an additional and efficient profit (Ranchordas, 2015) from an asset that, if not made available to the collaborative economy would not produce higher returns than those arising from a common lease agreement (Alfonso, 2016). When the owner assumes these costs, the price of the accommodation decreases, which allows the future guest to take this type of stay into consideration and weigh it according to his economic capacity. It should also be said that although the prices for tourists decrease, the prices for tenants rise, so much so that in Madrid and Barcelona, for example, in some neighborhoods, family homes were replaced by tourist houses (Gil & Fernandez, 2018).

State of the collaborative economy in the EU

As a growing phenomenon, the collaborative economy in the tourism industry started without much regulation (Vilalta Nicuesa, 2018), but the member states of the European Union began to see the advantages and the drawbacks of the interactions between the different actors in the collaborative economy arena. The EU established that the collaborative economy must be seen as a business opportunity because these platforms aim for the creation of 'digital value', which is a key concept for the future economic growth of the European Union and the consolidation of the digital single market, by means of ICT. Additionally, it is also pointed out that the appearance of the platforms has brought with it a change in the digital economy because they have the power to create and influence new markets and pose challenges that stimulate the adaptation of the traditional ones. According to the above, the dynamics of the platform creates new economic participants, diverse business lines and new business projects (COR, 2016).

The EU's attention to the collaborative economy has allowed it to issue several documents on how the member states should deal with this phenomenon, and has focused the academic literature on consumer rights (Dumancic, 2017; Petersen, 2018; Soltero Mariscal & Vargas-Hernández, 2017) and the contractual obligations of the platforms. In accordance with the above, the European Union has recognized the consumer relationship between the platform and the guest (COR, 2016) and has emphasized the importance of enhancing consumer protection in this arena, because many factors would appear to be advantageous to the platform, to the detriment of the consumer in an online scenario. These factors include: misleading advertising, the low or nonexistent legal responsibility of the platform when a dispute arises, little use of or insufficient information available on the option of using ADRs as an online dispute resolution platform (Rodriguez, 2017), misleading prices, forcing the consumer to accept foreign jurisdiction or applicable law to obtain the service and putting the consumer in the unfair position of having to sue for compensation in a country other than the one in which he/she resides.

Under the current state of affairs, in which we can see that the collaborative economy platforms have created a successful business without following the minimum legal parameters of consumer protection, the European Union has taken some corrective measures (EU, 2019) regarding the clauses established by these platforms. On the other hand, it is important to note that in the contractual relationships established between the platforms and the consumer-guests, the domicile of the parties, in most cases, guarantees the international nature of the contract, complicating the question of how a clause should be written and what characteristics it must have to be considered correct. The fact that a clause covering the terms and conditions subscribed between the consumer and the platform is in accordance with current consumer protection laws, means that the platforms have modified their services contracts, considering the consumer not only as a client but as a subject of special protection. The latter can be verified thanks to the intervention of the European Commission in the clauses of the Airbnb platform in July 2018.

The European Commission and the authorities for the protection of EU consumers urged the Airbnb platform to change the clauses contained in its terms and conditions of service, given that these failed to comply with several EU rules and regulations (European Commission, 2018), based on the lack of price transparency and the fact that the current clauses facilitated the exercising of other abusive business practices, harmful for the interests of the consumer. On the specific issue of international jurisdiction, the Commission indicated that Airbnb was misleading consumers when it established in its clauses as competent a jurisdiction different from that of the consumer's member state of residence. Additionally, Airbnb cannot unilaterally and without the consent of the consumer change the conditions of service because the consumer will remain in the dark regarding their new rights and obligations. In September 2018, the Commission announced that the platform had complied with what it was entrusted to do; leaving Airbnb terms and conditions more adapted to EU consumer protection regulations.

Methodology

Both qualitative-descriptive and qualitative-critical methodologies are used. Initially, the characteristics of the EU private international law regulations referring to international jurisdiction and applicable law are listed, as well as the consumer special rights contemplated in them. Finally, the qualitative-critical methodology is used to carry out a critical analysis of the clauses of four platforms: Toprural, Airbnb, Homelidays and Escapade to determine whether these clauses are drawn up in accordance with the EU consumer regulations and to ascertain their weaknesses.

In accordance with the above, the qualitative-critical methodology focuses on analyzing and pointing out the inconsistencies that arise between the text of the clauses and the legal requirements demanded by the European authorities for contracts involving consumers in order that changes can be introduced, with the aim of improving the wording or pointing out the missing legal elements so that the clause can be considered correct. In the special case of Airbnb, which has already been reviewed by the European Commission, it will be to verify if this platform continues to comply with the EU consumer protection law. The objective of using these methodologies is to answer the research question: are collaborative economy platforms in the tourism industry obeying the consumer law of the European Union?

Private international law regulations regarding the consumer

Private international law (PIL) is in case of a dispute between the parties, the regulation that is responsible for providing the answers on which court should have jurisdiction and which law it should apply to the contract, if the parties have failed to establish these points in their contractual agreements. It is important to note that, even if the parties have agreed on what specific law will be applied, there are circumstances in which these selections must be modified or added to, given that the rights of specially protected subjects such as the consumer, the employee or the policyholder and beneficiary of an insurance policy prevail, because EU legislation seeks to protect the weaker party to the contractual relationship (Rodriguez, 2014). In the EU, the rules of PIL are firstly contained in the EU regulatory instruments, only if we cannot apply them will the PIL give us the answer on the steps to follow in accordance with international conventions or treaties and finally it will refer us to the internal law of the country with jurisdiction.

In matters of international jurisdiction and applicable law, the EU has two regulations in force in civil and commercial matters and contractual obligations. Regulation (EU) No 1215/2012 of the European Parliament and of the Council of December 12 of 2012 on jurisdiction and the recognition and enforcement of judgments in civil and commercial matters (Brussels I bis Regulation, hereinafter, RBIbis) and Regulation (EC) No 593/2008 of the European Parliament and of the Council of June 17 of 2008 on the law applicable to contractual obligations (Rome I Regulation – RRI). However, it should be pointed out that the regulations are not directly applicable. When the parties do not reach an agreement on international jurisdiction and applicable law, we must first verify that the scope of each regulation is complied with, and after that, check whether the specific and independent concept of a consumer as defined under these rules is applicable the contractual party to which we intend to apply the rights and benefits contained in the regulations.

Thus, we have to first, with respect to RBIbis, check the four scopes of application: the material one, which determines whether the matter should be civil or mercantile, with some exceptions, article 1.2; the territorial, which states that the regulation is directly applicable in the member states territory, article 81.3; the cutoff date, which indicates that the matter must have been presented after 10 January 2015, article 81.2; and the scope of personal application or community connection, which determines which of the respondents' domiciles, located in an EU member state, will be considered the permanent residence, articles. 5-6. Regarding the RRI, only the scopes related to the material, territorial and cutoff date specifications are required, being aware that the latter applies to contracts concluded since 17 December 2009, these are dealt with in articles 28 and 29. Regarding the consumer, each regulation establishes special rights and an independent definition, but there are common characteristics that indicate that a consumer is an individual acting outside his trade or profession (Torres Lana, 2013), and although this definition is general and it is not related to the dynamics of the collaborative economy in the tourism industry, by defining the guest as a consumer we are including him/her within the spectrum of the protection of EU consumer regulations.

Analysis of the platform's clauses

In view of the above, let us begin with the analysis of the clauses of the four platforms to be reviewed, to determine if they are consistent with RBIbis and RRI; the platforms are Toprural, Airbnb, Homelidays and Escapade. It is essential to note that the clauses of these platforms are continually being modified, so it must be said that the text of the clauses discussed below was taken from the platforms' web sites on December 1, 2019. Additionally, it is important to consider that the clauses may change according to the country where the person who enters the website is located, due to the domain of the platforms as well as language, payment method allowed in the country, advertising, and the services that can be provided in that country. This short analysis was carried out from Spain. First, we have a jurisdiction and applicable law clause from Toprural, with a registered office in Madrid:

> *"Relations between the parties shall be governed exclusively by Spanish law. The resolution of any dispute between the Provider and the Traveler shall be the exclusive jurisdiction of the courts of the place of residence of the Traveler"* (Toprural, 2019).

In this brief clause, Spanish law is indicated as applicable law, which is correct under article 3 of the RRI (European Parliament, 2008) that states: "A contract shall be governed by the law chosen by the parties". This provision does not refer to the consumer rights stipulated under article 6 of the RRI, but merely establishes the applicable law without distinction between host and guest, since both participants can be considered as a 'party' in a legal conflict against the platform, whom it classifies as a provider. The last point is relevant because if the guest fulfills the requirements to be considered a consumer, then article 6 indicates that the law stipulated by the parties applies; however, the consumer does not lose the protection afforded by the rules of his/her country of primary residence.

With respect to the jurisdiction, the clause provides that the courts of residence of the guest are exclusively competent. This does not necessarily imply that the clause is in accordance with RBIbis (European Parliament, 2012) with regard to consumer rights provisions contained in articles 17, 18 and 19. First, we need to check article 19, and verify that this agreement does not fit into any of the three special cases provided for under this disposition, that we are going to explain later, and we move on to article 18, under which the consumer can sue the professional in either of two locations: his country of residence and the country in which the platform has its registered office.

Therefore, the Toprural clause determinates the exclusive jurisdiction of the courts of the guest's country of residence, the consumer, and does not contemplate the option that the consumer is entitled to have, of suing the platform in the jurisdiction where its registered office is located, as is established in article 18 of RBIbis, which stipulates that the consumer must have both options. As an example, we can say that a Belgian person who enters into a contract with Toprural, in case of dispute, can sue the platform in Belgium, where he/she has his/her domicile or in Spain, which is where the registered office of this platform is located, even if the clause establishes that the Belgian courts have exclusive jurisdiction. As a result of the above, we can say that this clause is incorrect because it lacks coherence with the provisions of the current consumer rights legislation established under RRI and RBIbis and could be considered null and void.

We should continue with the following platform, Airbnb, the most famous, profitable and competitive platform (Zervas et al., 2017) of the collaborative economy in the tourism industry, with double headquarters; the United States and Ireland. The popularity of Airbnb requires special consideration, before continuing with the analysis of international jurisdiction clauses and applicable law. Airbnb is the dominant platform in the current market and its success could be attributed to its initiative to invest in and develop a system of intermediation offering private homes as accommodation with economic benefits for all parties involved (Oskam & Boswijk, 2016). Additionally, Airbnb is the oldest for-profit platform of this nature that has managed to grow and evolve this system into a reliable business. Now, we can proceed to analyze the current international jurisdiction and applicable law clause displayed on the Airbnb website, which states the following:

> *"These Terms and Conditions will be governed and interpreted in accordance with Irish law. In the event that you act as a consumer, and the consumer defense regulations that you apply on a mandatory basis in your country of residence contain provisions that are more beneficial to you, those provisions will apply regardless of the choice of Irish law. In the event that you act as a consumer, you may initiate any legal proceedings in relation to these Terms and Conditions before the competent court of your place of residence or before the competent court of the registered office of Airbnb in Ireland. If Airbnb wishes to assert any of its rights against you as a consumer, you may only do so before the courts of the jurisdiction in which you are a resident. If you are acting as an entrepreneur, you agree to submit to the exclusive jurisdiction of the Irish courts."* (Airbnb, 2019)

In this clause, the law of Ireland is indicated as applicable, which is correct in the light of the RRI, because the parties can choose the law that will govern the contract without regard to the place where they are located. In addition, this clause in accordance with article 6 of the RRI refers to the protections afforded in a consumer contract, in which the consumer does not waive the rights granted in its own country of residence by accepting that the law of another state will be the applicable law. Regarding international jurisdiction, the clause contains very detailed information on the courts in which Airbnb can be sued, following the provisions of article 18 of the RBIbis, which stipulates that Airbnb can only sue the consumer at his/her domicile. However, the consumer has the option of either filing the lawsuit in the domicile of the platform or in his own. It is even stated that in case of acting as a professional or entrepreneur, the Irish courts will have jurisdiction, thus evidencing the special treatment that the consumer receives in comparison with the regular contractors.

In view of the above, and as the Commission has already pointed out, Airbnb contract clauses are now in compliance with consumer protection regulations. But it is important to say that the execution and compliance of the contract need a more exhaustive study with an analysis of specific cases to verify that the platform would respond to consumer requirements in a way that is consistent with the provisions established in its terms and conditions. Luckily, the Commission has already taken the first step to put this platform in check, this warning being the first indication of the need for special regulation to oversee platforms of this type.

Let us continue with the Homelidays platform (HomeAway, 2019), whose headquarters are located in London. Special consideration must be given to this platform, because it is part of the HomeAway group, which is a subsidiary of Expedia Group, Inc., whose registered office is in the United States. This platform has its registered office in England, which is still part of the 28 EU member states, and therefore the regulations apply directly to this platform. The question is whether the consumer can also sue the company Expedia Group, Inc., from within the boundaries of the Union, even if it is registered in a country outside the EU. The answer is found in article 17.2 of the RBIbis which states that the consumer can sue the professional service provider in a member state if he has a commercial establishment located in the EU, as is the case of Expedia Group, Inc., which, although it is in the United States, has a subsidiary in London, HomeAway. The RBIbis establishes that, in the case of the consumer, for all purposes, including jurisdiction, it will be considered that the professional with whom the service was hired is in a member state, if the company in question has a commercial establishment in the EU. Let us look at the international jurisdiction and applicable law clause viewable on the Homelidays platform:

> *"13.1 These Conditions applicable to the Traveler shall be governed by the laws of England and Wales and the English courts shall have non-exclusive jurisdiction to settle any claim arising from these conditions, without prejudice to the application of the laws and the jurisdiction of the place of Traveler's residence (if acting as a consumer). It should be noted, by way of example, that the mandatory provisions of Spanish legislation will apply, and the Spanish courts will be competent for Travelers residing in Spain"* (HomeAway, 2019).

In this clause, in the first place, it assigns non-exclusive jurisdiction to the English courts, which implies that although the parties establish specific courts to settle the dispute in case of conflict, they also decide not to forgo the option of filing the lawsuit in a forum different from the one agreed upon, which could be considered competent in accordance with the applicable norms for determining competence (Sánchez Lorenzo, 2012). Unlike the exclusive jurisdiction clauses that are applied directly, the non-exclusive jurisdiction clauses must be reviewed and analyzed by the judge, when performing the discretionary exercise in which he decides whether to decline (González, 2017) or accept the validity of this jurisdiction (Keyes, 2005). Non-exclusive jurisdiction agreements are provided for under the RBIbis, in article 25, which states that the jurisdiction chosen by the parties will be exclusive, unless they indicate otherwise.

Regarding the rights of the consumer under RBIbis, it is not valid to claim that the choice of non-exclusive jurisdiction is harmful to the consumer because although the parties choose the courts of a specific country as competent, the other applicable forums are still valid according to the parameters of the clause, even in a more in-depth study, this type of clause could be regulated by the exceptions applicable under article 19 of RBIbis. However, it should be said that the wording could more clearly establish the criteria of article 18 of RBIbis so that the consumer, who usually does not know his own rights, can more easily understand his options if he wishes to sue. Before we continue with the clause analysis, it is important to stop for a moment and explain the exceptions relating to international jurisdiction in matters of consumer rights.

As we can see in section 4 of the RBIbis, the agreements drawn up by the platforms in which the court of a specific country is established as competent when it has no relationship with the consumer may be valid if their content refers to the exceptions indicated in article 19 of the RBIbis. Thus, this article establishes three circumstances that allow the benefits of articles 17 and 18 not to apply, even if the contractual relationship is a consumer relationship. Article 19 indicates that these types of agreements will prevail; (i) if they are agreed after the initiation of the dispute between the professional and the consumer; (ii) if they allow the consumer to freely decide whether to file the lawsuit at their own domicile or at the professional's and the agreement establishes additional forums, allowing the consumer to voluntarily choose which one to go to; (iii) if both the professional and the consumer have their domicile or primary residence in the same member state at the time of entering into the contract, they can establish the courts of that State as competent, provided that it is permitted by the law of that State.

Once we have established the exceptions, we can continue with the analysis of the clause. On the example that the clause sets, it could be pointed out that there is an inconsistency between what the clause states in a general sense and what is indicated as an example. In the latter, it is indicated that, if you reside in Spain, the competent courts will be those of that country. Although one of the possibilities available under RBIbis is that the consumer can sue at his own domicile, there must always be the option for him to sue the professional in the state where he has his registered office, therefore, establishing a clause of this nature with a single forum, even if it includes the consumer's jurisdiction, is still wrong. Additionally, in the example, there is no reference to non-exclusive jurisdiction, which is extremely important in the drafting of the general clause. In the matter of applicable law, the wording of the clause complies with the provisions of the RRI, because, first, the laws of a specific state -England and Wales- are determined as applicable. Subsequently, the clause indicates that if a person acts as a consumer the laws of the traveler's place of residence still apply, taking into account that the traveler can be, the consumer-guest party.

Finally, we have the Escapade platform (Escapade, 2019) with headquarters in Girona, Spain. This platform has terms and conditions of service in which nothing is mentioned about international jurisdiction and applicable law. The question in this case would be whether Private International Law can help us find the answer when there is no agreement between the parties. First, it must be said that, since Spain is one of the EU members states, this platform, being in that country, must comply with EU regulations for consumer protection. In the case of international jurisdiction, the RBIbis articles 17, 18 and 19, relating to the jurisdiction in the matter of contracts concluded by consumers apply regardless of whether there is an agreement on which courts to go to in case of conflict, even when the parties reach agreements that are contrary to what is established in this article, the regulation still prevails.

Thus, if the parties do not agree on which courts would be competent in case of conflict, the consumer can sue the professional either at his own domicile or at the latter's. In the case of the professional, he can only sue the consumer at his own address. Regarding the applicable law, the RRI has a provision that defends the interests of consumers because it establishes that if there is no agreement between the parties, the contract will be governed by the law of the country in which the consumer has his primary residence. However, for the latter to be determined as applicable, one of the two conditions mentioned in the standard, concerning the business activities of the professional, must be met. The first, that this company pursues its commercial or professional activities in the country where the consumer has his primary residence, this being the natural market of the entrepreneur in question (Carrizo Aguado, 2018) and the second that, by whatever means, it directs such activities to that country or to several countries including that country, and the contract falls within the scope of such activities.

This concept of activities aimed at a member state should not be taken lightly. Establishing whether a professional directs his activities to a certain territory is not easy to determine when we talk about online platforms located anywhere in the world, which allow for the acquisition of goods or services through the internet. The Court of Justice of the European Union (CJEU) has indicated that there is evidence that can be used to determine this situation, revealing the international nature of the commercial activity, such as the listing of telephone numbers with the international prefix, the use of domains on the web pages of the seller that do not coincide with that of the member state where it is established; for the seller himself to claim to have clientele located in the member states; the establishment of payment in euros or in any of the currencies used in the member states and the choice of being able to change the currency; to allow users to change the language of the website to one of the languages commonly used in the EU, among others (CJEU, 2010)

This evidence alone does not prove that the commercial activities are directed towards a member state, these must be duly accredited in a trial, so that the judge can recognize the existence of a connection between the evidence indicated above and the commercial activities of the professional. The response of the PIL to the consumer that contracts services with the Escapade platform, which, as mentioned, does not have an international jurisdiction and applicable law clause, is clear. However, it has elements that must be analyzed in the light of the specific case, because if any one of the requirements that the regulation demands is not fulfilled, with respect to either the consumer or the professional, the consumer protection articles of the regulation could not be applied. In that case, it would be necessary to check if the contract fits in with the provisions of other articles by the nature of the parties involved or if not, to revise an international agreement or the internal or national law of the state where the parties intend to litigate.

Conclusion

As we can see, the answer to the question of whether or not the platforms of the collaborative economy we referred to are obeying the consumer law of the European Union is "no", in a quantitative way of speaking, because only one of the four platforms that were analyzed in matters of applicable law and jurisdiction is following RBIbis and RRI. It must be said that Airbnb is the only one that is obeying consumer rights regulations, doing so since 2018 because the European Commission gave them a deadline to fix their term and conditions and they complied, and they are continuing to do so; the other platforms have room for improvement. Therefore, the clauses listed by the platforms Toprural and Homelidays can be considered abusive, because there is a lack of transparency (Rodríguez Achútegui, 2016) in the terms of the agreement and because these are not consistent with EU consumer rights legislation, and favor the platform to the detriment of the consumer. With respect to the legal gaps generated by the absence of essential clauses in the terms and conditions of the Escapade platform, these can be filled with the RBIbis and RRI regulations.

The pattern here is that these platforms have detailed web pages regarding their services and the accommodation they offer, but in regard to their applicable law and jurisdiction clauses, they do not put much effort into investigating what is the correct way to establish the content of their terms and conditions according to European consumer law. As a result of this short investigation, we can say that three of the four platforms are currently in breach of EU regulations and only one of them continues to comply due to being the subject of special monitoring by the European Commission. In this context, the judge has been given the function of identifying and canceling abusive clauses that may violate the rights of consumers (Alonso Bezos, 2016). Additionally, to verify the possible nullity of these clauses, the content must also be verified, since these may be in flagrant violation of consumer rights (Hualde Manso, 2016), and therefore would be deemed null and void.

References

Airbnb (2019) www.airbnb.es. Available at: https://www.airbnb.es/terms. [Last Acess: 1 12 2019].

Alfonso, R. (2016) 'Aproximación jurídica a la economía colaborativa: diferentes realidades'. *Cuadernos de derecho y comercio*, 66, 13-15.

Alonso Bezos, J. (2016) 'Consumers and users. Personal loan. Unfair terms.' *Aranzadi Doctrinal Magazine*, 1, p. 1.

Botsman, R. & Rogers, R. (2010). *What's Mine is Yours: The rise of collaborative consumption. 1 ed.* Nueva York: Harper Business.

Carrizo Aguado, D. (2018) *Régimen jurídico de las operaciones internacionales de consumo en los servicios turísticos digitales*. Madrid: Dykinson.

CJEU (2010) As. C 585/08 y C 144/09, ECLI:EU:C:2010:740, of December 7 of 2010, Peter Pammer vs. Reederei Karl Schlüter GmbH & Co KG and Hotel Alpenhof GesmbH vs Oliver Heller. Luxemburgo: Court of Justice of the European Union.

COR (2016) *The Local and Regional Dimension of the Sharing Economy.* s.l.:European Committee of the Regions.

Dumancic, K. (2017). 'Collaborative economy - challenge for regulators', *Economic and Social Development,* 22, 647-648.

Escapade (2019) www.escapade.es. Available at: www.escapade.es/es/condiciones-generales;jsessionid=aaaG4qA5PoSuZ_eZ8tWax [Last Acess: 01/12/2019].

EU (2019) www.europa.eu. Available at: http://europa.eu/rapid/press-release_IP-18-5809_es.htm. [last access: 01 12 2019].

European Commission (2016) A European agenda for the collaborative economy. https://ec.europa.eu/growth/content/european-agenda-collaborative-economy-1_en

European Commission (2018) press release: "EU Consumer Protection Regulation: European Commission and EU Consumer Protection Authorities Urge Airbnb to Comply". s.l.:s.n.

European Parliament (2008) Regulation No. 593/2008 of the European Parliament and of the Council of 17 June 2008 on the law applicable to contractual obligations (Rome I). Rome: s.n.

European Parliament (2012) Regulation No. 1215/2012 of The European Parliament and of the Council of 12 December 2012 on jurisdiction and the recognition and enforcement of judgments in civil and commercial matters of applicable law (Brussels I bis). Brussels: s.n.

Fainmesser, I. (2014) *Exclusive Intermediation.* Working Paper 2014-3. Providence RI: Brown University, Department of Economics.

Zervas, G., Proserpio, D. and Byers J.W. (2017) The rise of the sharing economy: Estimating the impact of Airbnb on the hotel industry. *Journal of Marketing Research,* 54(5), pp. 688-690.

Gil, J. and Fernandez, J.S. (2018) The expansion of the tourist city and new resistances. The case of Airbnb in Madrid. *EMPIRIA. Journal of Methodology of Social Sciences,* 41, 15-32.

González, J. (2017) Declinatoria y Arbitraje Internacional. Revista de Derecho, *Empresa y Sociedad (REDS),* 10, 72.

HomeAway (2019) HomeAway. Available at: www.homeaway.es/info/quienes-somos/aviso-legal/terminos-condiciones-viajeros/ [last access: 1/12/2019].

Hualde Manso, T. (2016) Del consumidor informado al consumidor real. El futuro del derecho de consumo europeo.. 1 ed. Madrid: Dykinson.

Keyes, M. (2005) *Jurisdiction in International Litigation. 1 ed.* Sidney: The Federation Press.

O'Connell, J. F. & Williams, G. (2005) Passenger' perceptions of low cost airlines and full services: A case study involving Ryanair, Aer Lingus, Air Asia and Malaysia Airlines. *Journal of Air Transport Management,* 11(4), 260-263.

Oskam, J. & Boswijk, A. (2016) Airbnb: the future of networked hospitality businesses. *Journal of Tourism Futures,* 2(1), 27.

Petersen, C., Ulfbeck, V. & Hansen, O. (2018) Platforms as Private Governance Systems - The Example of Airbnb. *Nordic Journal of Commercial Law,* 1, 46-50.

Ranchordas, S. (2015) Does Sharing Mean Caring: Regulating Innovation in the Sharing Economy. *Minnesota Journal of Law, Science and Technology,* 16, 413.

Rodríguez Achútegui, E. (2016) The position of the Spanish courts with regard to the concept of consumer armed against unfair terms. *Aranzadi Doctrinal Magazine,* 5, 4.

Rodriguez, A. (2014) *Manual de Derecho Internacional Privado. 1 ed.* Madrid: Tecnos.

Rodriguez, I. (2017) El servicio de mediación electrónica y las plataformas de economía colaborativa. *Revista de Derecho Mercantil,* 305.

Sánchez Lorenzo, S. (2012). *Cláusulas en los contratos internacionales. Redacción y análisis. 1 ed.* Barcelona: Atelier.

Santolli, B. (2017) Winning the battle, Losing the war: European cities fight Airbnb. *George Washington International Law Review,* 49, 673-675.

Soltero Mariscal, D. & Vargas-Hernández, J. G. (2017) Modelos actuales de economías colaborativas Airbnb: cambiando la industria de la hospitalidad. *Lex Social,* 7(2), 353-355.

Toprural (2019) www.toprural.com. Available at: http://www.toprural.com/condiciones-de-uso-viajeros_l.html. [Last access: 01/12/2019].

Torres Lana, J. (2013) Derechos y garantías del contratante y del usuario de servicios turísticos adquiridos on line.. En: A. Paniza Fullana, ed. *Nuevas fórmulas de comercialización on line de servicios turísticos: subsunción en los tipos legales y distribución de responsabilidad.* s.l.:Comares, p. 22.

Vilalta Nicuesa, A. (2018) La regulación europea de las plataformas de intermediarios digitales en la era de la economía colaborativa. *Revista Crítica de Derecho Inmobiliario,* 275-278, 765.

6 An analysis of meal-sharing reviews to explore serendipity

Marina A. Petruzzi, Áurea Rodrigues, Michelle Moraes and Antonia Correia

The objectives of this chapter are to:

- ☐ Understand meal-sharing dynamics.
- ☐ Explore the dimensions that promote serendipity in meal-sharing experiences.
- ☐ Analyse online guest reviews through a content analysis technique.
- ☐ Present hospitality, food and tourist-tourist interaction as the most prominent dimensions of serendipity in meal-sharing experiences.
- ☐ Provide insights to improve marketing strategies of platforms and hosts.

Keywords: tourism, sharing economy, meal-sharing, serendipity

Introduction

In the period 2008-2018, the positive variation of tourism industry receipts was higher than the worldwide GDP growth (UNWTO, 2019a). In 2018, the European Union was the region with the highest total tourist receipts, while France and Italy were amongst the top six tourism earners in the world (UNWTO, 2019a). In the case of France, "consumer foodservice accelerated its digitalisation in 2018, which allowed for improved customer experiences and/or production optimisation. Whilst operators widely innovated in terms of digital tools to consolidate or gain share, digital usage varies significantly between channels" (Euromonitor International, 2019a: 45). The mentioned adaptation is not only related to the use of new technologies, but to the experience itself. In the Italian market, for example, "consumers are also showing themselves to be increasingly willing to try new products, ingredients and flavours" (Euromonitor International, 2019b: 33). Cross-cultural empirical studies confirm that novelty-change is a fundamental dimension inherent to innovation in food products (Guerrero et al., 2009). Another important aspect for travellers' food experiences is surprise, which was related to the simplicity, complexity and genuineness of these moments (Goolaup, Solér & Nunkoo, 2018).

In recent years, the number of innovative tourism experiences in terms of sharing economy initiatives has increased (WEF, 2019). Amongst the factors that influenced the growth of sharing economy after 2007 were the reduction of consumer trust in corporations and the purchasing power of consumers (European Union, 2013). In this context, some activities emerged and became key sectors in this area, such as home and car-sharing (Sigala, 2015), which is expected to present a revenue variation from USD 15 billion to USD 335 billion in the period 2014-2025 (UNWTO, 2019b). Concerning meal-sharing platforms, they are considered a potential market, which is currently underdeveloped (UNWTO, 2019b).

Conceptually, the sharing economy can be defined as "individuals offering their underutilized assets to others using digital platforms" (Bakker & Twining-Ward, 2018: 13). Thus, amid the aspects that differentiate sharing economy practices from traditional markets are the digital technologies that are used to match consumers and sellers, as well as the word of mouth reviews (Schor, 2014; WEF, 2017). Furthermore, the sharing economy is used to be related to eco-friendly initiatives, like the circular economy (OECD, 2019).

In tourism, the sharing economy represents a new way to increase the possibilities of connection between tourists and hosts (UNWTO, 2017): in these immersive experiences, tourists have the opportunity to enjoy a more local connection and interact with the hosts (Guttentag, 2015).

The expectation of unique experiences (Mao & Lyu, 2017) reinforces serendipity as an important enabler of sharing practices, including the meal-sharing sector. The term serendipity, which can be defined as the ability of making unexpected discoveries and finding interesting or valuable things by chance while travelling (Cary, 2004), has already been addressed by social scientists, medical humanists, and in applied research (Merton & Barber, 2006). Some of these studies have been conducted with organizations operating under the concept of the sharing economy. Mody, Suess and Lehto (2017) analyzed Airbnb and considered that the activity can create landscapes that offer serendipity. In turn, O'Regan and Choe (2017: 169), analyzing Airbnb and cultural capitalism, acknowledged that the new "market morality encourages tourists to seek the ease of market exchanges for experiences over serendipity". Boswijk (2017) suggested that Airbnb's success can be explained by a creative vision that resulted from serendipity. Although some research can be found in the context of the sharing economy, meal-sharing dynamics and serendipity can be considered an understudied subject.

By identifying this aspect, as well as the gap in the literature related to meal-sharing and serendipity, this chapter explores the dimensions that promote serendipity in meal-sharing experiences. To achieve this objective, a content analysis was performed through online reviews on the EatWith website, one of the main food service platforms (OECD, 2016). A deeper understanding of the main aspects that promote serendipity in sharing experiences can enhance the value created to consumers and sellers through food sharing experiences.

Literature review

Meal-sharing experiences and serendipity

Meal-sharing platforms have been designed to provide travellers with the possibility to connect with local residents and to engage in an overseas culture at the meal table (CNBC, 2015). This practice can become popular, similarly to other sharing economy practices such as home-sharing and car-sharing. Notwithstanding being widely discussed after the success of online platforms, the meal-sharing model was already addressed in the past as actions of eating a meal with family or drinking a beer with friends (Felson & Spaeth, 1978). The concept of meal-sharing relates to the supply of food and drink through online platforms connecting hosts to consumers looking for catering (UNWTO, 2017).

Meal-sharing organizations help residents to offer meal experiences to guests in their homes (Heo, 2016), becoming a way to connect individuals from all over the world. Furthermore, meal-sharing is recognized as offering personal and social interactions between individuals (Böcker & Meelen, 2017). Meal-sharing practices also bring benefits to those involved. For instance, it can provide jobs for people who need to work from home, generate less waste of food, promote social interactions (Sigala, 2017) and reinforce the local culture and traditions (Sims, 2009).

The practice of sharing meals involves more than just the sharing of food. It has the potential to reinforce social relations between the people who share the meal (Grignon, 2001; Sobal & Nelson, 2003), build communities and develop local social networks (Privitera, 2016). Meal-sharing can also be considered a form of engaging people in cooking and eating together, presenting a social essence (Marovelli, 2018). Hence, meal-sharing platforms allow tourists to immerse in a foreign culture during a meal at the residents' private homes (CNBC, 2015; Heo, 2016). Many platforms have emerged in the Internet era based on the sharing economy business model. In the sharing meal scenario, EatWith, EatWithalocal, Meal Sharing, and Cookening (Kim et al., 2018) are some examples.

Tourists opting for sharing economy activities may be looking for meaningful social interactions with locals and unique experiences in authentic settings (Böcker & Meelen, 2017; Ketter, 2019). Thus, when traveling and choosing a meal-sharing activity, tourists may be exploring new things and perhaps some of these experiences with others are more meaningful because they occur unplanned, becoming a serendipitous experience.

The term 'serendipity' seems to have been first used by Horace Walpole in 1754, denoting it as making discoveries, by accidents and sagacity (Merton & Barber, 2006). Over the years the meaning of serendipity has evolved, and a few authors have redefined it. Ryan (1991) for instance, considered serendipity to be the moment when the tourist is no longer a tourist and becomes subsumed by the context. Andel (1994) defined it as the ability to make an unsolicited discovery. Ferguson (1999) and Khan (1999) used serendipity to refer to happy accident and Tolson (2004: 51) as a "pleasant surprise". As such, serendipity not only refers to the surprise or the unexpected, but also to the ability of finding pleasure and meaning in accidental discoveries (Huang et al., 2014; Chandralal, Rindfleish & Valenzuela, 2015).

The serendipitous experience is unplanned but represents positive and memorable incidents (Chandralal et al., 2015; Tung & Ritchie, 2011) that can create unique opportunities for social interaction, build bonds among participants and generate connections with the place (Arsenault & Gale, 2004).

In the food context, an extraordinary experience must generate an element of surprise, followed by pleasure, which can be a result of the interaction with others or of the food (Goolaup et al., 2018). Furthermore, serendipity is suggested to positively influence the evaluation of the experience and consequently lead to satisfaction (Tung & Ritchie, 2011), making the experience more memorable, developing a sense of community (Mathiesen, 2012), improving the intentions to recommend (Chandralal & Valenzuela, 2013) and improving loyalty (Zhang, Wu & Buhalis, 2018). Furthermore, as serendipity is considered to play an important role in areas such as science, technology, and everyday life, more attention needs to be paid, as it can "offer unsought benefits in all these fields of endeavor" (Andel, 1994: 20).

Attributes and motivations of meal-sharing experience

Although this study does not focus exclusively on those tourists traveling specifically to experience gastronomic activities, some researchers describe this kind of activities as authentic, enriching and educational, where food is appreciated for the celebration of culture (Green & Dougherty, 2008; Tikkanen, 2007) and its experiential attributes (Sims, 2009). Researchers have analysed the themes that contribute to a memorable gastro-tourism experience and found seven dimensions: (1) deliberate and incidental gastro-tourists, (2) travel stages, (3) foodie risk-taking, (4) interdependent co-created tourist-host relationships, (5) authenticity, (6) sociability, and (7) emotions (Williams, Yuan & Williams, 2018).

Other researchers have explored the consumer demands for traditional and local food as a search for authenticity. Sims (2009) identified that most of the tourists in the UK had chosen a local food experience for the desire of authenticity, with products symbolizing the destination's culture and locality. The study argued that local food experiences can connect tourists to the region through its perceived culture and heritage. Böcker and Meelen (2017), when analyzing the motivation of people to participate in different forms of sharing economy in Amsterdam, identified that meal-sharing is powered by social motivations, with a strong social interaction component. The following sections highlight some of the key themes emerged from the literature considered more relevant for this study purposes.

Tourist-host relationship

Tourist-host social contact is stated to be a special form of cross-cultural contact (Fan et al., 2017). Cross-cultural social contact, interchangeably referred to as cross-cultural social interaction, is defined as the face-to-face contact between people from different cultural backgrounds (Yu & Lee, 2014). According to Bochner (1982), there are various branches of cross-cultural contacts according to the different criteria of classification defined, such as on whose territory the contact occurs, the time span of the interaction, the contact purpose, the type of involvement, the frequency of contact, the degree of intimacy between participants, relative status and power and numerical balance. In this way, 'eating local' translates the tourist's social integration into the host group (Bessière & Tibere, 2013). When experiencing tourist activities, the tourist is searching for the authenticity between the involved and not only the authenticity of the other (Wang, 1999). It seeks authenticity at the time of purchase – including food and drink – being a moment that does not make him/her feel that the experience is 'staged' but the interaction is spontaneous. This aspect makes him/her 'feel' like a local, developing authentic feeling through having specific products or services (Sims, 2009). In this context, the sociable interaction between host and guest in a food consumption environment brings the uncertainty of hosting someone who do not know (Ciborra, 1996), creating an experience that is recognized by some tourists as a source of pleasure that generates emotions (Kivela & Crotts, 2016). These emotions involved in the tourist activity can contribute to unforgettable experiences (Williams et al., 2018).

Food

According to Bessière (1998, p. 23) "food has symbolic characteristics that can appear in different guises":

i) food as a symbol – some foods are the basis of fantasy and concentrate symbolic virtues;
ii) food as a sign of communion – food shared and eaten with others;
iii) food as a class maker – food eaten in everyday life that can be translated into social class and/or lifestyles such as caviar or champagne;
iv) food as an emblem – this is the case with the culinary heritage of a given geographical area or community.

For Bjork and Kauppinen-Raisanene (2014), food constitutes a culture, which varies according to local history, socioeconomic and environmental conditions, food varieties and dietary preferences. Evidently, food and eating are linked to destinations. For Fischler (1993) the foodstuff is symbolically impregnated with values – perceived as positive or negative – which, incorporated by the eater, may influence his/her nature, behaviour and identity. The researcher developed the 'incorporation principle' in the eating behaviour, which is defined by "you are what you eat" (Fischler, 1988: 279).

On one hand, eating is the integration or adoption of qualities of the food that someone eats. In other words, it is the nutritional level (the food's nutrients are becoming our own body). On the other hand, "the person becomes part of a culture", that is, both food and cuisine, as they are culturally determined, place the eater in a social universe and in a cultural order (Crowther, 2018: 142). Eating habits are the basis of collective identity and, consequently, of otherness.

Tourist-tourist interaction

For Fanelli (2020) eating is a pivot around which our social lives revolve and feeding ourselves and others well is an essential part of socializing. Stone, Soulard, Migacz and Wolf (2018) found that meeting new people is one relevant element that leads to memorable food travel experience, as travel creates an environment to commune with and meet others. If eating some food makes one become more like that food, then those sharing the same food become more like each other (Fischler, 2011). Often memories derive from an environment where the traveller socialized with other tourists. The sociability is a form of association into which people engage for the pleasure of being with others (Bialski, 2012). It is a reciprocal way of interaction in which tact is very important and where sociability, in the form of casual conversation, is the finality itself (Lampinen & Ikkala, 2015). Sociability and authenticity are elements that should be present in any kind of tourism experience (Tussyadiah & Sigala, 2018).

Methodology

This study was exploratory in nature given the limited research with the literature concerning meal-sharing and serendipity. A netnography approach was adopted, based on online communities (Kozinets, 2010), which is becoming a popular method since an increasing number of consumers share their experiences online. The dataset is made of online reviews, considered an important source of data to understand tourists' experiences (Brochado, Troilo & Shah, 2017; Kozinets, 2010; Tussyadiah & Zach, 2017). The online reviews were gathered from open source information on the EatWith website (www.eatwith.com) in Italy and France. EatWith is considered the most potent brand among the global leaders' meal-sharing platforms (UNWTO, 2017). It describes itself as the world's largest community for authentic food experiences with locals and has around 5,000 culinary experiences available in over 130 countries (EatWith, 2018). Through the EatWith website, the tourist can choose the experience that he or she wants, from a cooking class to a rooftop dinner (EatWith, 2018).

The data collection followed three steps. First, we downloaded data from November 2013 to June 2019, from Italy and France. Both countries were chosen because they are important economies in the tourism industry, especially as gastronomic destinations (Hashimoto & Telfer, 2006). Second, we selected only the English reviews, to avoid translations that could be misunderstood, resulting in 3,876 reviews. To further identify serendipity characteristics, we searched for keywords in the context with both conceptual or denotative (as in the dictionary) and associated (as contextualized meaning) dimensions (Leech, 1981). The keywords used to the analysis were 'surprise', 'unexpected', 'wow', 'awesome', 'memorable', 'pleasure', 'delight', 'joy', 'enjoy', 'enjoyable', 'priceless' and 'rare'. These keywords were chosen because they are directly present in the definition of serendipity or represented by their synonyms.

A content analysis technique was used to identify the dimensions that promote serendipity in meal-sharing experiences. Content analysis is a qualitative technique of analysing written messages (Cole, 1988). This approach was selected as an accepted method of textual investigation, looking forward to revealing the enablers of serendipity in meal-sharing experiences. This study can be considered as exploratory and the data analysis was completed in the following steps (see Figure 6.1).

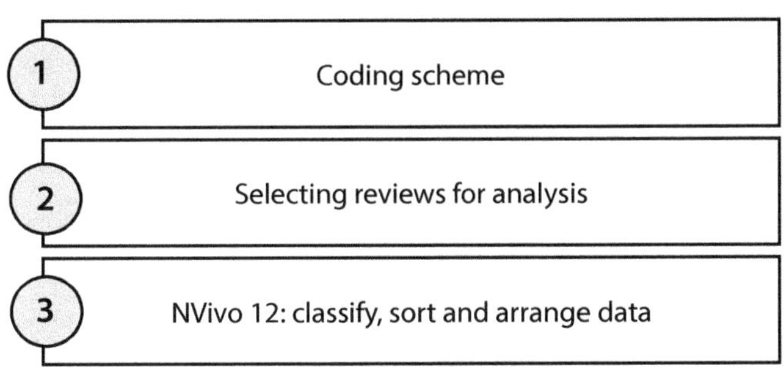

Figure 6.1: Content analysis steps. Source: Authors' elaboration

First, a deductive analysis was performed, with the categories and codes defined before the coding phase (Mayring, 2014). Following the concept of serendipity that emerged from the literature, the keywords representing the concept were used as the coding scheme. Each time that a keyword representative of the serendipity concept appeared, the review was selected for the analysis. Finally, NVivo 12, a qualitative research software, was used to help in the classifying process (QSR International, 2018). The software increases the rigor of the data analysis and reduces possible errors in the dissemination of data accompanied by manual analysis (Crowley, Harre, & Tagg, 2002).

Dimensions enabling serendipity in meal-sharing practices

In terms of word frequency related to the total reviews (3.876), it was possible to verify that both France and Italy reviews mentioned repeatedly the hosts and relational aspects such as conversation and company (Figures 6.2 and 6.3).

Figure 6.2: Word clouds, France. Source: Authors' elaboration

Figure 6.3: Word clouds, Italy. Source: Authors' elaboration

In terms of food and beverage, delicious and wine, as expected, were recurrent words both in France and Italy. Furthermore, the word clouds indicate the presence of the hosts' names in the reviews, who are part of the tourist-host interaction and therefore represent an aspect of serendipity in the 'hospitality' dimension. A good example is presented in the review "It was a complete surprise to all of us. Barbara is a fantastic host." Another review adds "Yves' talent, passion, and care in explaining his creations added priceless value to this exceptional dining experience". An aspect to be highlighted is the relevance of the guest's intention to recommend the experiences.

Through deeper analysis of the 3.876 reviews, 576 reviews containing the keywords representative of the concept of serendipity were identified. The content analysis carried out in this group of reviews allowed us to confirm that serendipity is much present in meal-sharing experiences. The qualitative content analysis acknowledged three main dimensions of meal-sharing practices that enable serendipitous experiences: hospitality, food, and tourist-tourist interaction. The most recurrent dimension in both France and Italy was hospitality, followed by food and tourist-tourist interaction (Table 6.1).

Table 6.1: Serendipity by main dimension by country

Dimension	France	Italy
Hospitality	54%	53%
Food	34%	35%
Tourist-tourist interaction	12%	12%

Source: Authors' elaboration.

In this context, the dimension 'hospitality' as an enabler of serendipity was described related to the tourist-host interaction, the pleasure of meeting the hosts, their family and pets, and related to the ambiance. In this sense, this dimension links serendipity with the pleasure and luck of interacting with the hosts, and the surprise of meeting with their family and pets. A good example of hospitality as a serendipity enabler is presented in the review "he welcomed us to the table where he joined us for dinner (an unexpected pleasure)". Another review expressed, "Loly (the host) is an absolute treasure". A third commented, "lucky us to have the pleasure to eat with Francoise". The 'hospitality' dimension also links serendipity with the atmosphere of the place in terms of the unexpectedness of the welcoming and the surprising ornamentation. A reviewer described this dimension as "amazing chef of colorful, artistic, healthy dishes, and a beautiful welcoming home full of interesting art and lots of colors". Another commented, "the ambiance was incredible, and the people were so friendly, we felt right at home!".

The 'food' dimension, in turn, as an enabler of serendipity, consists of reviews that suggest some sort of authenticity. It is related to the cultural and local ingredients, challenges and unique experiences as outcomes of the meal-sharing experience. The dimension is linked to serendipitous experiences resulting from the unexpected taste and the challenge that the food and flavours have incited. A typical review in this dimension expressed, "we loved the food, which challenged our palates and left us only in amazement of his creations". Another stated, "the food was unique, so fresh, and INSANELY delicious!". A third reviewer said, "never in my life have I had such an exquisite, authentic home cooked meal". A further addressed, "a delightful host who will also delight your palate with unexpected tastes. An excited reviewer shared, "it was unexpected... the blindfold added to the experience and it was fun talking to the other guests using your 'other' senses besides your sight!".

The interaction between tourists (the 'tourist-tourist interaction' dimension) was identified as an enabler of serendipity due to the fact that many reviewers expressed the positive surprise of having experienced the connection with other tourists, from all over the world, during the meal-sharing experience. This interaction was beyond their expectations, not being conceived as a plan in their pre-booking agenda. The dimension links serendipity to the unexpectedness of meeting other tourists, from different parts of the world and cultures, and the pleasure of meeting, chatting and sharing the meal with them. A typical review in this dimension expressed, "we had the pleasure of meeting wonderful people from around the world". Another review reported, "the unexpected bonus was developing friendships with such an eclectic group of people". Another shared, "absolutely a memorable way to experience Venice!! The food and fellowship with people from all around the world were outstanding!".

Finally, the three dimensions – hospitality, food and tourist-tourist interaction – are consistent with the notion of unexpected and surprising situations. The analyzed reviews suggest that situations in which tourists do not expect to face and/or that are beyond their expectations tend to result in memorable experiences and enduring memories in people's minds (Lynch & Srull, 1982; Talarico & Rubin, 2003), especially as they happen once (Cary, 2004).

Conclusion

The current study was undertaken in the context of the dimensions that promote serendipity in meal-sharing experiences, focusing on guests' reviews of the meal-sharing platform EatWith in France and Italy. This exploratory study, using a netnography approach, has demonstrated interesting dimensions emerging from the findings. This study not only identifies that serendipity is present in meal-sharing experiences but also suggests the dimensions of serendipity that influence meal-sharing experiences.

More specifically, the results indicating serendipity in meal-sharing experiences included three most prominent dimensions: the hospitality, the food and the tourist-tourist interaction. These dimensions were developed through investigation of guests' reviews of meal-sharing practices and linkages between existing literature on food consumption and serendipitous experiences.

From a theoretical perspective, this chapter has explored the concept of serendipity in meal-sharing experiences, which appears to be the first time that this approach has been taken. From a practical perspective, the findings from this research may give insights to platform operators to provide training to hosts when creating offers. Additionally, providers can improve the experiential aspects of their offer through the enhancement of serendipity inducers. Given the fact that meal-sharing is becoming an expressive phenomenon, this study has contributed a valuable understanding of the dimensions that provide serendipity in meal-sharing experiences, which can be used to improve the marketing strategies of platforms and hosts. Furthermore, it is important to highlight the relevance of this kind of strategy to the current economic system reconfiguration, which is towards more eco-friendly initiatives.

Despite the contributions this first approach has made to the context of meal-sharing and serendipity, it is not without limitations. First, this study focused exclusively on the EatWith platform, which is one of the most potent brands, although not the only one. Second, it is a qualitative exploratory study. Future studies could develop and apply a questionnaire with the three most prominent dimensions evidenced by this chapter – hospitality, food and tourist-tourist interaction – to verify our results through quantitative methodologies. Furthermore, future studies could focus on comparing our findings with the configurations present in other platforms related to sharing economy. Another area of study could also address the similitudes and/or differences of serendipity enablers between meal-sharing practices and traditional restaurants.

References

Andel, P. V. (1994). 'Anatomy of the unsought finding. serendipity: Orgin, history, domains, traditions, appearances, patterns and programmability'. *The British Journal for the Philosophy of Science*, 45(2), 631-648.

Arsenault, N. & Gale, T. (2004). *Defining Tomorrow's Tourism Product: Packaging experiences*, Department of Tourism and Culture, Yukon Government.

Bakker, M. H. E., & Twining-Ward, L. D. (2018). *Tourism and the Sharing Economy: Policy & Potential of Sustainable Peer-to-Peer Accommodation.* Washington, DC: World Bank Group.

Bessière, J. (1998). 'Local development and heritage: traditional food and cuisine as tourist attractions in rural areas'. *Sociologia Ruralis*, 38(1), 22-32.

Bessière, J. & Tibere, L. (2013). 'Traditional food and tourism: French tourist experience and food heritage in rural spaces'. *Journal of the Science of Food and Agriculture*, 93(14), 3420-3425.

Bialski, P. (2012). 'Technologies of hospitality: How planned encounters develop between strangers'. *Hospitality & Society*, 1(3), 245–260.

Bjork, P. & Kauppinen-Raisanen, H. (2014). 'Culinary-Gastronomic tourism – a search for local food experiences'. *Nutrition & Food Science*, 44(4), 294-309.

Bochner, S. (Ed.). (1982). *Cultures in Contact: Studies in cross-cultural interaction (Vol. 1).* Elsevier.

Böcker, L. & Meelen, T. (2017). 'Sharing for people, planet or profit? Analysing motivations for intended sharing economy participation'. E*nvironmental Innovation and Societal Transitions*, 23, 28-39.

Boswijk, A. (2017). 'Transforming business value through digitalized networks: A case study on the value drivers of Airbnb'. *Journal of Creating Value*, 3(1), 104-114.

Brochado, A., Troilo, M. & Shah, A. (2017). 'Airbnb customer experience: Evidence of convergence across three countries'. *Annals of Tourism Research*, 63, 210–212.

Cary, S. H. (2004). 'The tourist moment'. *Annals of Tourism Research*, 31(1), 61-77.

Chandralal, L. & Valenzuela, F. R. (2013). 'Exploring memorable tourism experiences: Antecedents and behavioural outcomes'. *Journal of Economics, Business and Management*, 1(2), 177-181.

Chandralal, L., Rindfleish, J. & Valenzuela, F. (2015). 'An application of travel blog narratives to explore memorable tourism experiences'. *Asia Pacific Journal of Tourism Research*, 20(6), 680-693.

Ciborra, C. (1996). 'Introduction: What does groupware mean for organizations hosting it?' In C. U. Ciborra (Ed.), *Groupware and Teamwork: Invisible Aid or Technical Hindrance,* John Wiley, pp. 1–19.

Cole, F. L. (1988). 'Content analysis: process and application'. *Clinical Nurse Specialist*, 2(1), 53-57.

Consumer News and Business Channel CNBC (2015). 'Sharing Economy - What's cooking in the sharing economy?' Retrieved June 21, 2019, from https://www.cnbc.com/2015/09/16/sharing-economy-plenty-of-meal-sharing-platforms-emerge.html.

Crowley, C., Harré, R. & Tagg, C. (2002). 'Qualitative research and computing: methodological issues and practices in using QSR NVivo and NUD*IST'. *International Journal of Social Research Methodology*, 5(3), 193-197.

Crowther, G. (2018). *Eating Culture, An anthropological guide to food (2nd Ed),* Toronto: University of Toronto Press.

Eatwith. (2018). Eatwith - about us. Retrieved June 24, 2019, from https://www.eatwith.com/pages/about-us.

Euromonitor International (2019a). Consumer Foodservice in France.

Euromonitor International (2019b). Consumer Foodservice in Italy.

European Union (2013). *The Sharing Economy: Accessibility Based Business Models for Peer-to-Peer Markets*. European Union.

Fan, D., Zhang, H., Jenkins, C. & Lin, P. (2017). 'Does tourist-host social contact reduce perceived cultural distance?' *Journal of Travel Research,* 56(8), 998-1010.

Fanelli, R. (2020). 'Seeking gastronomic, healthy and social experiences in Tuscan agritourism facilities'. *Social Sciences*, 9(2), 1-15.

Felson, M. & Spaeth, J.L. (1978), 'Community structure and collaborative consumption: a routine activity approach', *American Behavioral Scientist,* 21(4), 614-24.

Ferguson, A. (1999). 'The lost land of serendip'. *Forbes*, 193-194.

Fischler, C. (1988). 'Food, self and identity', *Social Science Information*, 27(2), 275-292.

Fischler, C. (1993). 'A nutritional cacophony or the crisis of food selection in affluent societies', in P. Leathwood, M. Horisberger and W.P. James (eds.), *For a Better Nutrition in the 21st Century*, New York: Vevey/Raven Press, pp. 57-65.

Fischler, C. (2011). 'Commensality, society, and culture'. *Social Science Information*, 50(3-4), 528–548.

Goolaup, S., Solér, C. & Nunkoo, R. (2018). 'Developing a theory of surprise from travelers' extraordinary food experiences'. *Journal of Travel Research*, 57(2), 218-231.

Green, G. P. & Dougherty, M. L. (2008). 'Localizing linkages for food and tourism: culinary tourism as a community development strategy'. *Community Development*, 39(3), 148–158.

Grignon, C. (2001). 'Commensality and social morphology: an essay of typology'. In P. Scholliers (Ed.), *Food, Drink and Identity Cooking, Eating and Drinking in Europe Since the Middle Ages* (pp. 23–33). New York: Berg.

Guerrero, L., Guàrdia, M. D., Xicola, J., Verbeke, W., Vanhonacker, F., Zakowska-Biemans, S. ... & Scalvedi, M. L. (2009). 'Consumer-driven definition of traditional food products and innovation in traditional foods. A qualitative cross-cultural study'. *Appetite*, 52(2), 345-354.

Guttentag, D. (2015). 'Airbnb: disruptive innovation and the rise of an informal tourism accommodation sector'. *Current Issues in Tourism*, 18(12), 1192–1217.

Hashimoto, A. & Telfer, D. J. (2006). 'Selling Canadian culinary tourism: Branding the global and the regional product'. *Tourism Geographies*, 8(1), 31-55.

Heo, C. Y. (2016). 'Sharing economy and prospects in tourism research'. *Annals of Tourism Research*, 58, 166–170.

Huang, W. J., Norman, W. C., Hallo, J. C., Mcgehee, N. G., Mcgee, J. & Goetcheus, C. L. (2014). 'Serendipity and independent travel'. *Tourism Recreation Research*, 39(2), 169-183.

Ketter, E. (2019). 'Eating with EatWith: analysing tourism-sharing economy consumers'. *Current Issues in Tourism*, 22(9), 1062-1075.

Khan, M. H. 1999. 'Quest for serendipity'. *The (Bangladesh) Independence*, June 15.

Kim, S., Lee, K. Y., Koo, C. & Yang, S. B. (2018). 'Examining the influencing factors of intention to share accommodations in online hospitality exchange networks'. *Journal of Travel & Tourism Marketing*, 35(1), 16-31.

Kivela, J. & Crotts, J. C. (2006). 'Tourism and gastronomy: gastronomy's influence on how tourists experience a destination'. *Journal of Hospitality and Tourism Research*, 30(3), 354–377.

Kozinets, R. V. (2010). *Netnography: Doing ethnographic research online*. London: Sage.

Lampinen, A. & Ikkala, T. (2015). Monetizing Network Hospitality: Hospitality and Sociability in the Context of Airbnb. Cscw, 1033–1044. DOI: 0.1145/2675133.2675274.

Leech, G. N. (1981). *Semantics*. Hardsworth: Penguin.

Lynch Jr, J. G. & Srull, T. K. (1982). 'Memory and attentional factors in consumer choice: Concepts and research methods'. *Journal of Consumer Research*, 9(1), 18-37.

Mayring, P. (2014). *Qualitative Content Analysis: theoretical foundation, basic procedures and software solution*. Sage, Klagenfurt.

Mao, Z. & Lyu, J. (2017). 'Why travelers use Airbnb again?' *International Journal of Contemporary Hospitality Management*, 29(9), 2464–2482.

Marovelli, B. (2018). 'Cooking and eating together in London: Food sharing initiatives as collective spaces of encounter'. *Geoforum*, 99, 190-201.

Mathiesen, L. (2012). 'The exploration of the memorable tourist experience', in J. S. Chen (ed.) *Advances in Hospitality and Leisure* (Vol. 8), pp. 21-43, UK: Emerald Pub.

Merton, R.K & Barber, E (2006). *The Travels and Adventures of Serendipity: A Study in Sociological Semantics and the Sociology of Science*. Princeton University Press.

Mody, M.A., Suess, C. & Lehto, X. (2017). 'The accommodation experiencescape: a comparative assessment of hotels and Airbnb', *International Journal of Contemporary Hospitality Management*, 29 (9), 2377-2404.

Organization for Economic Cooperation and Development OECD (2016), 'Policies for the tourism sharing economy', in *OECD Tourism Trends and Policies 2016*, OECD Publishing, Paris.

Organization for Economic Cooperation and Development OECD (2019), *Business Models for the Circular Economy: Opportunities and Challenges for Policy*, OECD Publishing, Paris.

O'Regan, M. & Choe, J. (2017). 'Airbnb and cultural capitalism: Enclosure and control within the sharing economy'. *Anatolia*, 28(2), 163-172.

Privitera, D. (2016). 'Describing the collaborative economy: Forms of food sharing initiatives'. In *Economic Science for Rural Development Conference Proceedings* 43, 92-98.

QSR International. (2018). 'What is NVivo?' Retrieved October 20, 2018, from https://www.qsrinternational.com/nvivo/what-is-nvivo.

Ryan, C. (1991). *Recreational Tourism: A social science perspective*. Routledge.

Schor, J. (2014). 'Debating the sharing economy. Great Transition Initiative'. Retrieved October, 20, 2018, from http://www.greattransition.org/publication/debating-the-sharing-economy.

Sigala, M. (2015). 'From demand elasticity to market plasticity: A market approach for developing revenue management strategies in tourism'. *Journal of Travel and Tourism Marketing*, 32(7), 812–834.

Sigala, M. (2017). 'Collaborative commerce in tourism: implications for research and industry'. *Current Issues in Tourism*, 20(4), 346-355.

Sims, R. (2009). 'Food, place and authenticity: local food and the sustainable tourism experience'. *Journal of Sustainable Tourism*, 17(3), 321-336.

Sobal, J. & Nelson, M. K. (2003). 'Commensal eating patterns: A community study'. *Appetite*, 41(2), 181–190.

Stone, M., Soulard, J., Migacz, S. & Wolf, E. (2018). 'Elements of memorable food, drink and culinary tourism experiences'. *Journal of Travel Research*, 57(8), 1121-1132.

Talarico, J. M. & Rubin, D. C. (2003). 'Confidence, not consistency, characterizes flashbulb memories'. *Psychological Science*, 14(5), 455-461.

Tikkanen, I. (2007). 'Maslow's hierarchy and food tourism in Finland: five cases'. *British Food Journal*, 109(9), 721–734.

Tolson, J. (2004). 'A word's eventful journey'. *US News & World Report*, 136(4), 51-51.

Tung, V. & Ritchie, B. (2011). 'Exploring the essence of memorable tourism experiences'. *Annals of Tourism Research*, 38(4), 1367-1386.

Tussyadiah, I. P. & Sigala, M. (2018). 'Shareable tourism: tourism marketing in the sharing economy'. *Journal of Travel and Tourism Marketing*, 35(1), 1–4.

Tussyadiah, I. P., & Zach, F. (2017). 'Identifying salient attributes of peer-to-peer accommodation experience'. *Journal of Travel and Tourism Marketing*, 34(5), 636–652.

Wang, N. (1999). 'Rethinking authenticity in tourism experience'. *Annals of Tourism Research*, 2(1), 349–370.

World Economic Forum WEF (2017). Collaboration in cities: From sharing to 'sharing economy', Available on: http://www3.weforum.org/docs/White_Paper_Collaboration_in_Cities_report_2017.pdf.

World Economic Forum WEF (2019). 4 big trends for the sharing economy in 2019. Available on: https://www.weforum.org/agenda/2019/01/sharing-economy/.

World Tourism Organization UNWTO (2017). *Affiliate Members Report, Volume 16 – Second Global Report on Gastronomy Tourism*. Madrid.

World Tourism Organization UNWTO (2019a). *International Tourism Highlights*, UNWTO, Madrid.

World Tourism Organization UNWTO (2019b), *New Business Models in the Accommodation Industry – Benchmarking of Rules and Regulations in the Short-term Rental Market*, Executive Summary, UNWTO, Madrid.

Williams, H. A., Yuan, J. & Williams Jr, R. L. (2018). 'Attributes of memorable gastro-tourists' experiences'. *Journal of Hospitality & Tourism Research*, 1–22.

Yu, J. & Lee, T. (2014). 'Impact of tourists' intercultural interactions'. *Journal of Travel Research*, 53 (2), 225–38.

Zhang, H., Wu, Y. & Buhalis, D. (2018). 'A model of perceived image, memorable tourism experiences and revisit intention'. *Journal of Destination Marketing & Management*, 8, 326-336.

7 Consumer perception of service quality: The case of Airbnb and Couchsurfing

Marie Dewitte, Jérôme Mallargé and Alain Decrop

The objectives of this chapter are to:

- ☐ Provide a better understanding of the service evaluation process in peer-to-peer accommodation services.
- ☐ Highlight differences in the assessment of service quality between two types of platform: Airbnb and Couchsurfing.
- ☐ Challenge the scale's dimensions used to assess the service quality in the sharing economy (i.e., CC-QUAL).
- ☐ Recommend a holistic assessment of service quality in evaluating together offline and online interactions.
- ☐ Highlight the differences between commercial and non-commercial platforms in terms of expected roles and governance norms (institutional governance vs. community-based governance).

Keywords: AirBnB; Couchsurfing; peer-to-peer accommodation; sharing economy; service quality; service evaluation

Introduction

Recent economic, social and environmental concerns have drawn attention to the necessity to rethink our consumption patterns (Barnes & Mattsson, 2016) and call for alternative forms of consumption. In parallel, digitalization dramatically changes the way we live, work, consume and travel (OECD, 2020). As a result, new consumption practices have emerged in the last years, privileging access over ownership (Botsman & Rogers, 2010). Those practices, labelled as sharing economy or collaborative consumption (Belk, 2014; Benoit et al., 2017; Botsman & Rogers, 2010), involve most of the time peer-to-peer exchanges (for a fee or for free) that are coordinated through community-based online services (Hamari, Sjoklint & Ukkonen, 2016). Such collaborative services have recently boomed, impacting many sectors, including the hospitality and tourism industry (Sigala, 2017), with well-known initiatives such as Airbnb or Couchsurfing.

Peer-to-peer accommodation services are transforming the tourism industry (PWC, 2015) by enabling consumers to share and access goods escaping traditional services like hotels and travel agencies. According to Hotrec (2014), peer-to-peer accommodation is twice bigger than the conventional tourism accommodation industry in Europe. The World Bank Group estimates a 31% annual growth of this new accommodation type between 2013 and 2025, which is six times bigger than the annual growth of the conventional bed and breakfast and hotel industry. In total, peer-to-peer accommodation makes up about 7% of accommodation worldwide (Bakker & Twining-Ward, 2018).

Two of the most sucessful sharing economy unicorns, AirBnB and Couchsurfing, have very different business models. Airbnb is an online peer-to-peer marketplace that matches hosts wishing to share their home with travelers (i.e. guests) who are looking for accommodation. Valued at 38 billion USD (Forbes, 2018), Airbnb has more than 60 million customers and around two million accommodations in the world (OECD, 2016). At the opposite, Couchsurfing is a free online hospitality exchange network that connects travelers looking for a place to sleep with people offering their 'couch' for a couple of nights. The community gathers around ten million members around the world.

Driven by the idea of new, original and authentic experiences, travelers turn more and more to such collaborative initiatives with the aim of sharing moments of intimacy and individuating their experience as far as possible (Bialski, 2012; Steylaerts & Dubhghaill, 2012). Those platforms also respond to the traditional industry limitation, as for instance high transaction costs, distorted information between market actors, asymmetrical regulations, and impersonal tourism experiences (Dredge & Gyimóthy, 2015). Although these factors contribute to explain the emergence of collaborative tourism, little is known about how consumers evaluate such collaborative services. However, this question is crucial given the challenges involved in this new type of exchange.

The first challenge is related to the peer-to-peer interactions (Botsman & Rogers, 2010). As such, consumers are co-creating the service, blurring the frontier between the consumer and the peer provider. The second challenge is related to the platforms and the market logic they follow. Indeed, peer-to-peer exchanges are often mediated or facilitated by an online platform that can be monetized or not. The monetization of the service makes the whole expectation and experience different from one collaborative initiative to another. Monetization also impacts the nature of hospitality. This nature can be conditional (i.e. involving automatic reciprocity) or unconditional (i.e. involving no reciprocity or generalized reciprocity) depending on the monetization of the exchange (Ikkala & Lampinen, 2015). Also, depending on the conditional or unconditional nature of hospitality, peer-to-peer accommodation services are positioned between pure sharing – a non-reciprocal prosocial behavior (Benkler, 2004) – and pure market exchange, which is driven by a business logic inducing reciprocity and monetary compensation (Belk, 2014).

This chapter seeks to better understand the service evaluation process in the particular case of the peer-to-peer accommodation sector. We investigate how differences between commercial initiatives, such as Airbnb, and non-commercial initiatives, such as Couchsurfing, may influence the consumer evaluation. Based on an analysis of online archival data and narratives, the study provides insights about how the type of intermediary and the specificities of each initiative affect perceived service quality assessment.

Context and related work

Facilitated by information technologies and Web 2.0 (Belk, 2014; Hamari, Sjöklint & Ukkonen, 2016) that enable the transformation of online interactions into offline experiences, "collaborative consumption is the latest addition to numerous developments and trends in the marketplace that have substantially transformed traveler behavior and disrupted the industry dynamics" (Tussyadiah & Pesonen, 2016: 1025). According to the most common definition, collaborative consumption is "an economic model based on sharing, swapping, trading or renting products and services enabling access over ownership" (Botsman & Rogers, 2010: xv). In a more recent definition, Benoit et al. (2017) define collaborative consumption as a market-based relationship involving three actors, a platform, a peer service provider, and a peer consumer in which ownership transfer is excluded. This definition suggests that collaborative consumption differs from traditional forms of exchange on three dimensions: 1) the number and type of stakeholders, 2) the nature of the exchange, and 3) the directness of exchange.

The first challenge concerns the triadic aspect of the relation. The presence of three actors in the exchange makes the relationship more complex (Eckhardt et al., 2019; Liu et al., 2019). This difficulty is even bigger in the case of service failure (Mallargé, Decrop & Zidda, 2019). Indeed, consumers are no longer passive, and co-create the service in a more active way (Huang & Benyoucef, 2013;). Second, because of the peer-to-peer nature of the exchange, consumers have potentially an 'enhanced role' because they could be providers as well (Eckhardt et al., 2019; Harmari, Sjoklint & Ukkonen, 2016; Narasimhan et al., 2018). In this triadic relation, the platform role is often to facilitate the transaction between consumers and providers (i.e. in the case of tourism, hosts and guests) (Benoit et al., 2017, Perren & Kozinets, 2018).

Concerning the nature of the exchange, collaborative consumption favors access over ownership. Consumers can access a good (e.g. a room or an apartment) for a limited period of time and usually for an lower price (or even for free in some cases, as for Couchsurfing). Consumers' motivations in peer-to-peer accommodations are mainly utilitarian (e.g. economic), hedonic (e.g. enjoyment, quest for authenticity) and social/symbolic (e.g. group/community belonging; for a review, see Decrop et al., 2018). In particular, the social dimension of sharing is present in the hospitality and tourism industry as consumers are going to meet other people, to share a common place, and to learn from one another, developing a real sense of community.

This sense of community is enhanced by hospitality networks, which refer to the way consumers are connecting to each other using social networking systems. On the one hand, consumers connect with one another through the platform and build trust, which is a key condition of collaborative consumption (Botsman & Rogers, 2010; Bardhi & Eckhardt, 2012;

Möhlmann, 2015). These online interactions allow them to share personal information with the aim of getting to know each other better before meeting or sharing a common place. The sharing of information allows consumers to connect with others who look like themselves and share the same interests (Finley, 2013; Botsman et al., 2010). The sense of community is reinforced through the creation of interpersonal links and a larger network (Belk, 2007; Perren & Kozinet, 2018) leading sometimes to a kind of pseudo-kinship (Belk, 2001, 2007, 2010).

On the other hand, hospitality networks also involve face-to-face interactions that are emotionally intense (Molz, 2012). Hospitality emphasizes the importance of the relationship between host and guest (Brotheron, 1999) and goes beyond material aspects (Lampinen, 2015). We can differentiate collaborative initiatives based on whether there is a conditional or unconditional hospitality (Ikkala & Lampinen, 2015). Conditional hospitality implies a monetary compensation like in Airbnb. In that case, the exchange is assimilated to market exchange. The monetary nature of the exchange increases consumers' expectations about the service as they see it, more like a traditional, market-based exchange. Market-based exchanges are defined by Belk (2014) as "pseudo sharing" as these induce a negative reciprocity (Sahlins, 1972) in which people give in the hope of receiving back. In contrast, if hospitality is unconditional, it is free without any expectation of reciprocity, like in Couchsurfing. In that case, the exchange is closer to pure sharing. Consumers are motivated by meeting new people, learning from their culture and living authentic moments. Social interaction matters more than the free accommodation.

Platforms play a crucial role in anticipating and preparing for this interaction, as they help to generate trust between strangers. However, platforms such as Airbnb or Couchsurfing also develop tools such as reviews, ratings and comments to help people trust one another. Although the aim is to reduce perceived risks and uncertainty, review and rating systems play a double role. First, for new travelers, previous comments help them shape their expectations. Second, review and rating systems allow consumers to evaluate concretely the experience in setting scores and to list positive and negative points. Nevertheless, literature shows that rating systems are neither perfect nor very reliable (Filippas, Horton & Golden, 2018). That is also probably why communities develop outside the platforms via blogs or forums. However, all those media remain an interesting starting point to assess the quality of peer-to-peer accommodations.

Service evaluation and perceived service quality

Perceived quality can be defined as "the consumer's judgment about an entity's overall excellence or superiority" (Zeithaml, 1987 in Parasuraman, Zeithaml & Berry, 1988: 15). Parasuraman, Zeithaml and Berry describe perceived quality as an overall evaluation similar to attitude. Similarly to satisfaction, service quality is the result of a comparison between expected and perceived performances (Grönroos 1982, 1984). Nevertheless, satisfaction is more related to a specific transaction, whereas service quality is a global judgement. According to the SERVQUAL scale, the assessment of service quality includes five dimensions: tangibles, responsiveness, assurance, empathy and reliability (Parasuraman, Zeithaml & Berry, 1988).

The rise of the Internet and the development of online services call for a new service quality assessment scale, taking pure online players and the specificities of online shopping into account. E-S-QUAL (Parasuraman, Zeithaml & Malhotra, 2005) measures online service quality and consists of four dimensions: efficiency, fulfilment, system availability and privacy. The emergence of collaborative services has made service quality evaluation further evolve. Today, online and offline relationships coexist in a three-way game. The evaluation of the service concerns the platform as well as the peer provider. Therefore, Marimon et al. (2019) created a scale based on those previous seminal works. In addition to the classical process related to the scale development, they consulted collaborative platforms' CEOs to validate their items selection. In doing so, this scale named CC-QUAL attempts to address some of the gaps resulting from the specificity of collaborative services. SERVQUAL, E-S-QUAL and CC-QUAL's dimensions are summarized in Table 7.1.

Few studies so far have investigated the service evaluation process in collaborative consumption. Möhlmann (2015) and Tussyadiah (2016) research the main determinants of satisfaction whereas Yang et al. (2017) focus on commitment and loyalty. These researchers all use a quantitative approach, as opposed to Mallargé, Decrop and Zidda (2019) who use a qualitative approach to identify several sources of (dis)satisfaction and question previous researches in their use of the classical disconfirmation paradigm. They show that the disconfirmation paradigm as well as the two-factor theory help in understanding the service evaluation process and they highlight differences between market and non-market exchanges.

Table 7.1: Dimensions of the three quality scales

CC-QUAL	
Dimensions	**Description**
Site organization	Design of the site that makes it appealing and easy to browse
Platform responsiveness	Quickness to deal with and to establish agreements
Legal protection & trustworthiness	Privacy and legal protection, reliability and honesty of the published information
Peer service provider	Professionalism, honesty and empathy of the peer service provider
Social Interaction	Interaction experience with people (including other users and the peer provider)
Hedonics*	Characteristics that make interaction with the site pleasant and enjoyable
Tangibles*	Assessment of the physical appearance of the shared assets
SERVQUAL	
Dimensions	**Description**
Tangibles	Physical facilities, equipment, and appearance of personnel
Reliability	Ability to perform the promised service dependably and accurately
Responsiveness	Willingness to help customers and provide prompt service
Assurance	Knowledge and courtesy of employees and their ability to inspire trust and confidence
Empathy	Caring, individualized attention the firm provides its customers
E-S-QUAL	
Dimensions	**Description**
Efficiency	The ease and speed of accessing and using the site
Fulfillment	The extent to which the site's promises about order delivery and item availability
System availability	The correct technical functioning of the site
Privacy	The degree to which the site is safe and protects customer information

**Dimensions not retained for the final scale*

Source: adapted from Marimon et al (2019); Parasuraman et al. (1988, 2005).

In the same perspective, our study focuses on the specific context of collaborative tourism. Our research goes a step ahead in the process of investigating the perceived service quality of peer-to-peer accommodation services. More specifically, we focus on the particular case of Couchsurfing and AirBnB because of their different business models. We consider how the unique features of collaborative exchanges, and especially triadic relationships, affect the assessment of service quality. To do so, we build on previous works on service quality and we confront our findings with the dimensions identified in SERVQUAL, E-S-QUAL and CC-QUAL scales.

Methodology

A qualitative exploratory approach has been chosen in order to understand the service quality assessment. Such an approach is likely to generate unique insights about how consumers behave in an open and indirect way (Belk, Fischer & Kozinets, 2013). We combined two methods for data collection. First, we proceeded to an analysis of archival online data publicly available from the Airbnb and the Couchsurfing websites. Online qualitative research contributes to a deeper understanding of a phenomenon in a naturalistic and unobtrusive way (Kozinets, 2006). We analyzed thematically more than 800 posts in English and in French from forums and blogs, describing the customer experience. We only considered messages posted after 2010 to ensure recent information.

Next, in order to triangulate our data, a series of 23 incidents reporting satisfying or unsatisfying experiences on Airbnb and Couchsurfing were collected and analyzed according to the critical incidents technique (Flanagan, 1954). Any incident should meet four criteria for being included in our sample (Bitner et al, 1990): (a) relate a sharing tourism experience, (b) be very satisfying or dissatisfying, (c) be discrete and (d) be sufficiently detailed to be pictured by the researcher. It is worth mentioning that this empirical study is part of a bigger project investigating the three major sectors of the sharing economy: accommodation, transportation and goods. For this study, we only focus on the tourism accommodation sector.

Results

There are significant differences in the way travelers evaluate the service, depending on the initiative. Differences are seen during the interactions with the platform, when a service failure occurs and during the interactions with the peer service provider.

Online interactions – the role of trust and reliability

In peer-to-peer accommodation services, travelers mainly use the platform with the aim at finding an accommodation. Data reveal differences in the way people search for information and how they evaluate that information when choosing an accommodation. Both Couchsurfing and AirBnB websites offer much information, such as the host's profile, description and pictures. To enhance trust between peers, they also use reputation systems. Comments, ratings and references are analyzed by travelers to reduce uncertainty in their selection of the best host. What differentiates both initiatives is the kind of information travelers consider to make their choice and the elements they take into account to evaluate the service.

On Couchsurfing, travelers focus on two main elements: the safety and the host's profile and interests. Travelers are concerned by safety because they are sharing a place with someone they do not know. The safety issue is mainly mentioned by female hosts or guests.

> *"Of course general precaution is necessary especially if you are surfing. Make sure you see the references and a vouched and verified member is always better… Also take time to go through your host's profile and make sure you have some common interests, this way you have greater chances of enjoying your stay."* – Beatgeneration

> *"I only host solo females for safety reasons and I first want to know if we have something in common. I want to have the idea that we have a good chance to get along and have nice conversations. In Couchsurfing, the idea is that the traveler searches for hosts they think they may click with, for instance because of shared interests or passion."* – 8-23-2016ta-j

Sharing common interests is central in the Couchsurfing community. It helps people connect more easily with each other and make sure that they will spend pleasant moments together sharing about their passions. That is why Couchsurfers value long profile descriptions. They can learn about the person they will meet, their interests and what type of people they get along with. Couchsurfers pay attention to photos to assess the person's reliability. Similarly, hosts accept people who share the same interests and who take the time to send a personalized message explaining why they want to connect with them.

> *"P.S: a complete profile (photo, full description) is more accessed and gets more positive answers. A 'bond' between you and every person you meet is created (you can give a quick comment on the other's profile) and it helps other couchsurfers to understand who you really are."* – Cyril

Profiles, photos and comments are even more valued on AirBnB but travelers also focus on the characteristics of the accommodation rather than on the interests of the host.

> *"I recommend you to read the comments and reviews of former guests who did already book the place. It gives you a good idea of the apartment and of the host and helps you with the final choice."* – ThomasMimi

"We wanted that the apartment was conforming to the photos and it was the case. I think that being honest regarding the photos you put on AirBnB is really important. If there is no match between the photos and the reality, it gives a first bad impression and the trust is broken." – Henry

Travelers mainly pay attention to the reviews and examine the photos to check whether the accommodation is clean and in good state. They expect the housing unit to look like the photos on the advertisement.

"I had a bad experience with AirBnB as the place I rented was misrepresented. Very far from center of town when listing said 5 minutes. We were going to stay 5 days, left after 1 night. Place also had ants, no soap or any paper towels, napkins or TP. Furniture and mattress were awful". – Sally G

In that sense, travelers value the host's honesty and they try to assess the trustworthiness of the platform. They make sure the host is verified by the platform because they do not want to be a victim of fraud, like ChristineA39.

"In June '12, I used AirBnB to book an apartment for two weeks in Brussels with a host named Maximillian. When we arrived exhausted at 6 am, he said our original apartment had water damage and took us to an apartment in another neighborhood with almost none of the amenities that he listed. And almost no furniture. That's right, no place to put anything. No glasses to drink water from! Not even curtains on the windows. […]. To make matters worse, AirBnB will not allow me to leave a review to warn others because I cancelled my reservation with Max and left after 4 days. Even though I still had to pay for those days. It's a flaw with the AirBnB and one that everyone should be aware of." – ChristineA39

This story reveals that service failure and the way the failure is handled is a major element of the service quality evaluation.

Service failure – who is in charge?

When bad experiences happen, travelers question the trustworthiness of the photos and the reliability of the host. To avoid such incidents, travelers carefully read the comments of former travelers, check the host's answer rate and cancellation rate, and pay attention to the cancellation policy of the platform. They firstly seek for information to assess the advertisements' reliability and then ask other users for 'tricks' to avoid bad experiences. Data reveal that the failure attribution process seems to be different depending on the initiative. On AirBnB, the service failure is often attributed to the online platform. Travelers expect the third party to solve their problems and to compensate for the failure.

AirBnB

"We booked an AirBnB in Montreal last April and it was a disaster!! AirBnB doesn't provide any guarantee on the apartment quality, you have to pay in advance, you can't reach someone in case of trouble. Phone calls to California ruined us. We were isolated and without accommodation. We paid €390 for a nice and shiny apartment and received a dirty, old and degraded apartment. We finally got refund €300 with a lot of difficulties. AirBnB is over for us". – Niko34090

Indeed, Niko34090 considers that AirBnB is accountable for the service failure and the recovery. Most of AirBnB users expect a contact person to give guarantees in case of problems. They expect to be helped by the platform and to get a refund.

"We have used AirBnB extensively in the past, but unfortunately they have 'dropped the ball'. They allow misleading advertisements and do not vet what hosts advertise. We are currently in Oristano Italy in a tiny cramped room, nothing like what was advertised! AirBnB are slack in not checking what is advertised on their site and you need to question in detail before booking." – PJL50

Likewise, travelers consider that AirBnB is responsible for controlling the hosts on the platform and protecting their data. It suggests a misconception of what AirBnB is and how it works as a peer-to-peer exchange platform. At the opposite, some AirBnB users consider that the platform is reliable and is not accountable for the service failure, simply being an intermediary.

"You can't blame the platform. You must blame the host if he cancelled your reservation. The website can, unfortunately, not force the host to honor the contract since it is just an intermediary." – Ti_Fleur_Fanée

This perspective shared by some travelers also stresses the importance of all the guarantees offered by the platform to protect customers, including the verification of hosts, the cancellation policy and delayed payment. In that sense, the platform protects consumers and gives sufficient guarantees.

"You pay the rent to AirBnB and AirBnB only pays the host when you are in the apartment. If there is a cancellation in the meantime, AirBnB pays you back quickly and gives you some compensation." – ThomasMimi

Couchsurfing

For Couchsurfing, most service failures occur when the host and the guest do not share the 'ethos' or mindset of the Couchsurfing community.

> *"You need to know that Couchsurfing is a 'state of mind'. Surfing on someone's couch isn't the same that staying at a hotel!! You have to be ready for improvisation, to adapt yourself, be open-minded and most of all be ready to share some time with your host."* – Cocci47

Couchsurfers insist on the need to participate in the exchange to really experience Couchsurfing. Travelers value social interactions more than the free bed and expect the host to be open to discussion and to give some of his time.

> *"During my trip to Island with a friend, we used Couchsurfing and stayed two nights at Reykjavik. It was great to have a free accommodation but I regret that the host didn't spend more time with us. We just spoke for one or two hours on the entire stay. It would have been nice that she shares a meal with us and that we speak more about our respective lives."* - Oliver

From the host perspective, meeting new people and new cultures is also important. Couchsurfing should be a two-way process, as emphasized by Mariha2912:

> *"I had a conversation with a few friends a couple of months ago. One of them used to host people in Athens. Not anymore. He said that 70% of people were arrogant, using his place only as a place to sleep for free, not acknowledging him, not trying to make a conversation or get to know him, often been demanding and leaving the house in a mess. And many refused to offer accommodation in return, using all kinds of excuses when he was visiting their part of the world. Couchsurfing is supposed to be for people traveling on the cheap, aiming to meet new people and explore new cultures, being open-minded and willing to spend time with their host, building some sort of bond. Focus is supposed to be on the whole experience, not just a free bed."* – Mariha2912

The host expects the guest to participate in cooking, in doing the groceries or at least to show some appreciation of what is offered. It is also important for the travelers to follow the house's rules and to behave in a respectful way. If it is not the case, the service failure is attributed to the guest who does not fit with the philosophy of the experience and does not want to participate into the community. The platform is not included in the evaluation as it is a simple intermediary.

The quote of Mariha2912 also illustrates the central role of offline interactions in the service quality evaluation. Results illustrate the different weights given to social versus functional criteria, depending on the initiative.

Offline interactions – functional versus social criteria

The social dimension appears to be central in Couchsurfing. The sense of community is expressed in a lot of comments written by Couchsurfers and really differentiates the two initiatives. The wish to meet new people and discover another culture is widespread in the Couchsurfing community who see travel as a life experience. Interactions with local communities, meeting new people, discussing with them, making friends or understanding the culture are valued by Couchsurfers. Accordingly, Couchsurfers evaluate the service based on those social criteria.

> *"I've been to Warsow with my boyfriend for a 4-days trip. We stayed at Macej's place. He hosted us on the sofa-bed of his living room. We had our sleeping bag with us. The first day, he gave us his apartment's keys so we were free to move when we wanted. He also gave us a map with places to visit and walked with us in the city to show the most important curiosities and explain a bit about the history of the city. During the following days, we were independent. We spent one evening playing board games with him but for the rest of the time, he had to work on his PhD. The last evening, we were supposed to go out together but since it was raining, he made us a typical Polish dish with a glass of apple vodka (too strong for me). We couldn't refuse it! So far, this evening remains one of the best of my life. Macej had lots of friends, made us discover Warsow's night life and many other things an ordinary tourist wouldn't have discovered."* - Diane

As mentioned in Diane's story, having the host sharing privileged moments with her was important. She valued visiting the city with a local and discovering things she would not have discovered otherwise. Travelers engaging in Couchsurfing know that they will not always sleep in nice rooms with all the required comfort but they will live a more authentic experience and share moments of intimacy.

In contrast, tangibles and functional aspects are turning out to be crucial for travelers using AirBnB.

> *"I came back from a trip in New-York and I booked my apartment via AirBnB. The apartment was conforming to the description: 50m^2, fully equipped kitchen, air conditioning, garden, phone/television/internet, nicely*

decorated. The owner was friendly and helpful. He offered a barbecue when we arrived and gave us a lift to the airport when we left. The neighborhood was calm and 30 minutes by bus from Time Square. The reviews of former travelers were trustworthy. I paid 1057 USD for 4 people for 15 days, it was perfect for me." – Ledzep68

AirBnB travelers evaluate the service based on functional criteria such as equipment, price, ease of use, convenience etc. In his story, Ledzep68 mainly stresses the functional aspects of the experience and the fact that the rented apartment was in line with the description. Nevertheless, he also values the fact that the owner was friendly and helpful. As illustrated in the previous quote, the social dimension is generally limited to common courtesy and sympathy. AirBnB users are very sensitive to the social aspect of the exchange but they do not expect the same level of interactions than in Couchsurfing. They consider mainly AirBnB as a cheaper accommodation solution. They are happy to meet lovely and helpful people but in contrast with Couchsurfers, they neither look for a sense of belonging to a travel community nor living authentic moments with their hosts. They nonetheless are very attentive to the host's honesty and professionalism, as mentioned before. However, when such high degree of socialities appears, it is lived as a bonus as the story of Céline illustrates:

"In the summer 2015, I went to Croatia with my best friend for a roadtrip. On our way home, we stayed at a AirBnB in Germany and that's so far the best experience of my life with AirBnB. We stayed at a gay couple and we were received like queens. [...] We had dinner together and they did insist to prepare the meal for us and make us taste specialties. We spent the night discussing about our holiday and get acquainted. To be honnest, I don't expected as much when I use AirBnB. I want to meet local people and have real cultural exchanges. And it's less and less the case with AirBnB. [...] I just exepted that the room was consistent with the ad on AirBnB. I wasn't expecting much more because if you want interactions with the locals, you chose for Couchsurfing. I just expected from the host to be available to give me the keys and be nice." - Céline

Although not central in the evaluation, the social dimension can improve the experience, surprise users and even delight them as illustrated by Céline's story. We summarize the main findings in Table 7.2.

Table 7.2: Summary of findings about service evaluation – A comparison of Airbnb and Couchsurfing

	Distinctive elements			
CC-QUAL's dimensions	Applicability	AirBnB	Applicability	Couchsurfing
Site organisation	X		X	
Platform responsiveness	V	Payment security Accountability in case of service failure	X	
Legal protection	V	Reputation system and guarantees reduce uncertainty	X	
Trustworthiness*	V	Focus on tangibles and fit with description	V	Focus on safety and common interests
Peer service provider	V	Peer is seen like a market professional	V	Peer is seen as a friend
Social interactions	V	Honesty and professionalism are expected	V	Sharing privileged moments and personalized experiences is valued
Tangibles**	V	Cleanliness, price, convenience and location	X	
Hedonics**	X		V	Online community supports the general sense of community

*merged with legal protection in the final CC-QUAL scale

**not retained for the final CC-QUAL scale

Source: Authors' own elaboration.

Discussion

This study compares the two best known collaborative initiatives in the hospitality industry, i.e., AirBnB and Couchsurfing, that are opposed in terms of business models. In this comparison, we highlight significant differences in the way travelers evaluate the service quality in both initiatives, especially concerning dimensions taken into account in their service quality assessment. We use CC-QUAL as a framework for our analysis.

Based on the SERVQUAL scale (Parasuraman, Zeithaml & Berry, 1988), measuring consumers' perceptions of service quality and the E-S-QUAL scale (Parasuraman, Zeithaml & Malhotra, 2005), assessing electronic service quality, CC-QUAL (Marimon et al., 2019) was developed for collaborative consumption services. The scale includes five dimensions to assess the quality of the service created by the three agents involved: the customer, the peer service provider and the platform. First, 'site organization', 'platform responsiveness and agility' and 'legal protection and trustworthiness' are related to the interaction with the website. Then 'peer service provider' and 'social interactions' are related to the interaction between peers. 'Tangibles' and 'hedonics' were first considered by the authors but eventually deleted from the scale. Our results actually support the CC-QUAL scale dimensions and challenge them in the way they are structured. We also highlight the limitation of the scale, which is not applicable in all context.

Regarding the interaction with the platform, the two dimensions of the CC-QUAL scale ('platform responsiveness and agility' and 'legal protection and trustworthiness') contribute well in the service quality evaluation. First, we showed that in the case of a service failure, AirBnB users expect a service recovery from the platform. Thus, service quality evaluation includes the capacity of the website to manage customers' complaints and offer guarantees in case of a problem, which refers to the responsiveness and agility dimension of the CC-QUAL scale. This dimension is not observed among Couchsurfing users who do not expect any recovery from the website in case of bad experiences. This difference can be explained by the commercial grounding of the initiatives. AirBnB is a for-profit organization whereas Couchsurfing is a non-profit one. The monetization of the service leads people to categorize it as a traditional service and conceptualize the platform as a conventional service provider inducing reciprocity and monetary compensation (Belk, 2014). In contrast, Couchsurfers consider the platform as a simple 'matchmaker' between hosts and guests. In that sense, we contribute to the literature in responding to the call of Benoit et al. (2017: 226) "to investigate who customers blame for service failures and who is expected to recover for the failure as well as the form of recovery that will be required". We also show that governance norms are different. In AirBnB, consumers clearly value strong institutional and legal governance mechanisms (Akbar & Tracogna, 2018; Perren & Kozinets, 2018, Wirtz et al., 2019), whereas Couchsurfing is rather ruled by 'community-based governance' (Adler, 2001). AirBnB users appreciate dealing with 'verified' hosts and expect AirBnB to only allow trustworthy hosts on the platform.

Second, both AirBnB and Couchsurfing users spend a lot of time and attention on the reputation system to choose the service provider. It allows them to decrease uncertainty about the outcome and prevent them from bad experiences, which refers to the legal protection and trustworthiness dimension of the CC-QUAL scale. The difference between the two initiatives lies in the elements travelers focus on. AirBnB users focus more on tangible and functional elements such as the apartment cleanliness, the furniture and the location, and expect the accommodation to look like in the advertisement. Those elements are criteria on which they will evaluate the service once consumed. Here, we clearly highlight a limitation of the CC-QUAL scale. It removes the 'tangibles' dimension from the scale although it seems to be a crucial factor in perceived service quality assessment. In Couchsurfing, travelers are concerned by their own safety and use the reputation system to choose hosts who have well-established profiles and with whom they share common interests. They want to be sure they will share insightful moments with the host. This capacity to share special moments with the host and to build some kinds of bond is also used as criteria for service evaluation. We thus see that online and offline interactions are interdependent. Travelers use the platform to make sure that both functional expectations and social expectations will be met.

Regarding the peer service provider attributes (i.e. honesty, professionalism and empathy) and the social interaction, our results illustrate a significant difference in the evaluation process between Couchsurfing and AirBnB. First, it highlights a different perception of the service provider's role. When AirBnB users evaluate the service based on tangible and functional criteria such as price, convenience, cleanliness and location, Couchsurfers value social interactions and the shared experience Couchsurfing offers. The role of the peer service provider is thus defined differently whether the service is monetized or not. In AirBnB, travelers expect the peer service provider to be nice, honest and 'professional'. They value spending time with the host when entering the accommodation but do not expect to share personalized moments with the host. Because AirBnB is driven by a business objective and relates to market-exchange, consumers' expectations toward the peer provider are similar to the ones they could have in a traditional service. This is probably one of the elements explaining why it is nearly impossible to create a feeling of belonging and a feeling of community in such access-based model (Bardhi & Eckhardt, 2012; Eckhardt et al., 2019). On the other hand, Couchsurfing, before being defined as a peer-to-peer accommodation service, is seen as a hospitality network and a community (Decrop et al., 2018). Indeed, Couchsurfers do not focus on functional aspects of the service and do not mind sleeping on the floor but they really count on the host to spend time with them, make them visit the city and share memorable experiences.

Second, our results reflect the importance of the community in the hospitality sector, especially for the non-profit organizations. This dimension however does not appear in the final CC-QUAL scale. The online community and the wish for travelers to analyze comments and exchange online was mentioned in the CC-QUAL scale under the 'hedonic' label. This confusing label (that only concerns online interactions) was retrieved from the final scale by CC entrepreneurs involved in the study. This is nonetheless an important element in the hospitality industry as it fosters trust between stranger peers. Besides, the 'sense of community' and the face-to-face interactions that are the foundations of Couchsurfing do not clearly appear in the scale. Similarly, the importance of tangibles in Airbnb does not appear in the final CC QUAL scale as well. Nevertheless, it seems to be an important dimension in the evaluation process for some collaborative services. Our study thus highlights a second limitation of the CC-QUAL scale, as it cannot be generalized to every sector of the collaborative consumption. Finally, it also highlights that the pseudo-sharing rationale of AirBnB opposed to the pure sharing logic of Couchsurfing show that governance norms differ.

Our results thus bring empirical support to the dimensions of the CC-QUAL in the sense that both the platform and the peer service-provider participate to the service quality evaluation. But we also identify significant divergences. The CC-QUAL scale investigates the platform's perceived quality and the peer's service perceived quality separately, while we suggest that perceived quality cannot be evaluated using online and offline interactions independently. While dimensions such as trustworthiness are considered separately in online and offline perspectives, our results show that they are connected. Indeed, consumers use online cues to assess the host's trustworthiness and assess the online platform's trustworthiness by the trustworthiness of the service provider it registers. Service quality must be evaluated globally, taking offline and online encounters together.

Finally, our study contributes to Perren and Kozinets' work (2018) by differentiating for-profit and non-profit organizations. Perren and Kozinets conceptualize collaborative markets through the idea of 'Lateral Exchange Markets' and propose a classification of four distinct market configurations. These market configurations are defined according to the degree of platform intermediation, on the one hand, and the degree of consociality between actors – defined as the actors' physical or virtual copresence and the opportunity for social interactions – on the other hand. Couchsurfing and AirBnB platforms are both 'matchmakers' as they mediate a service flow between hosts and guests and are characterized by high levels of platform intermediation and of consociality. The degree of platform mediation and consociality nonetheless differs between the two initiatives. Platform mediation is higher in Airbnb as it coordinates and regulates the communication between peers, the payments and the reputation system. It gives assurance to the users and coordinates the pairing of hosts and guests. In Couchsurfing, platform mediation is less important in the sense that it allows people to connect and only provides a reputation system. It offers few tools to formally manage the exchange, relying more on the users' community and mutual trust. Levels of consociality are also different. Consociality is higher in Couchsurfing as it is based on an online and offline community. Couchsurfers can not only connect to share accommodation but also to go have a drink, visit the city or meet for a specific event with locals. Social interactions lie at the core of the initiative. In Airbnb, consociality is comparatively less prominent but the platform also favors social interactions between peers and offers tools for travelers to meet and discuss before the service is experienced.

References

Adler, P.S. (2001) 'Market, hierarchy, and trust: The knowledge economy and the future of capitalism', *Organization Science*, 12 (2), 215-234.

Akbar, Y.H. & Tracogna, A. (2018) 'The sharing economy and the future of the hotel industry: transaction cost theory and platform economics', *International Journal of Hospitality Management*, 71, 91-101.

Bakker, M. & Twining-Ward, L. (2018) *Tourism and the Sharing Economy: Policy and potential of sustainable peer-to-peer accommodation*, World Bank.

Bardhi, F. & Eckhardt G. M. (2012) 'Access-based consumption: the case of car sharing', *Journal of Consumer Research*, 39 (4), 881–98.

Barnes, S. J. & Mattsson, J. (2016) 'Understanding current and future issues in collaborative consumption: A four-stage Delphi study', *Technological Forecasting and Social Change*, 104, 200-211.

Belk, R. W. (2001) *Collecting in a Consumer Society*. Routledge.

Belk, R. (2007) 'Why not share rather than own?', *The Annals of the American Academy of Political and Social Science*, 611 (1), 126-140.

Belk, R. (2010) 'Sharing', *Journal of Consumer Research*, 36 (5), 715-734.

Belk, R. (2014) 'You are what you can access: Sharing and collaborative consumption online', *Journal of Business Research*, 67 (8), 1595–600.

Belk, R., Fischer E. & Kozinets R. V. (2013) *Qualitative Consumer and Marketing Research*, Thousand Oaks, CA: Sage.

Benkler, Y. (2004) 'Sharing nicely: On shareable goods and the emergence of sharing as a modality of economic production', *Yale Law Journal*, 114, 273.

Benoit, S., Baker, T. L., Bolton, R. N., Gruber, T. & Kandampully, J. (2017) 'A triadic framework for collaborative consumption (CC): Motives, activities and resources & capabilities of actors', *Journal of Business Research*, 79, 219-227.

Bialski, P. (2018) 'Home for hire: How the sharing economy commoditises our private sphere', in Anthony Ince and Sarah M. Hall (ed), *Sharing Economies in Times of Crisis: Practices, Politics and Possibilities*, New York, NY: Routledge, pp. 83–95.

Bitner, M. J. (1990) 'Evaluating service encounters: the effects of physical surroundings and employee responses', *Journal of Marketing*, 54 (2), 69-82.

Botsman, R. & Rogers R. (2010) *What's Mine Is Yours: The Rise of Collaborative Consumption*, New York, NY: HarperBusiness.

Decrop, A., Del Chiappa, G., Mallargé, J. & Zidda, P. (2018) '"Couchsurfing has made me a better person and the world a better place": The transformative power of collaborative tourism experiences', *Journal of Travel and Tourism Marketing*, 35 (1), 57–72.

Dredge, D. & Gyimóthy, S. (2015) 'The collaborative economy and tourism: Critical perspectives, questionable claims and silenced voices', *Tourism Recreation Research*, 40 (3), 286-302.

Eckhardt, G. M., Houston, M. B., Jiang, B., Lamberton, C., Rindfleisch, A. & Zervas, G. (2019) 'Marketing in the sharing economy', *Journal of Marketing*, 83 (5), 5-27.

Filippas, A., Horton, J. J. & Golden, J. (2018) 'Reputation inflation'. In *Proceedings of the 2018 ACM Conference on Economics and Computation*. New York: Association for Computation Machinery, 483-484.

Finley, K. (2013) *Trust in the Sharing Economy: An Exploratory Study*, Centre for Cultural Policy Studies, University of Warwick

Flanagan, J. C. (1954) 'The Critical Incident Technique', *Psychological Bulletin*, 51 (4), 327–58.

Forbes (2018) 'As a rare profitable unicorn, Airbnb appears to be worth at least $38 billion', viewed 14 June 2020, www.forbes.com/sites/greatspeculations/2018/05/11/as-a-rare-profitable-unicorn-airbnb-appears-to-be-worth-at-least-38-billion/

Grönroos, C. (1993) 'An applied service marketing theory', *European Journal of Marketing*, 16 (7), 30-41.

Hamari, J., Sjöklint, M. & Ukkonen, A. (2016) 'The sharing economy: Why people participate in collaborative consumption', *Journal of the Association for Information Science and Technology*, 67 (9), 2047-2059.

HOTREC (2014) 'HOTREC position vis-à-vis the emerging shadow hospitality economy in Europe', 17 October, Riga.

Huang, Z. & Benyoucef, M. (2013) 'From e-commerce to social commerce: A close look at design features', *Electronic Commerce Research and Applications*, 12 (4), 246-259.

Ikkala, T. & Lampinen, A. (2015) 'Monetizing network hospitality: Hospitality and sociability in the context of Airbnb', paper presented at the *18th ACM Conference on Computer Supported Cooperative Work & Social Computing*, 14-18 March, Vancouver.

Kozinets, R. V. (2006) *Netnography. Handbook of Qualitative Research Methods in Marketing*, London, UK: Sage.

Lampinen, A. (2015) 'Deceptively simple: Unpacking the notion of "sharing"'. *Social Media + Society*, 1 (1).

Liu, P. J., Lamberton, C., Bettman, J. R. & Fitzsimons, G. J. (2019) 'Delicate snowflakes and broken bonds: a conceptualization of consumption-based offense', *Journal of Consumer Research*, 45 (6), 1164-1193.

Mallargé, J., Decrop, A. & Zidda, P. (2019) 'Revisiting satisfaction with collaborative exchanges in the sharing economy', in R.W. Belk (ed.) *Handbook of the Sharing Economy*, Edward Elgar Publishing, pp.303-315

Marimon, F., Llach, J., Alonso-Almeida, M. & Mas-Machuca, M. (2019) 'CC-Qual: A holistic scale to assess customer perceptions of service quality of collaborative consumption services', *International Journal of Information Management*, 49, 130-141.

Möhlmann, M. (2015) 'Collaborative consumption: Determinants of satisfaction and the likelihood of using a sharing economy option again', *Journal of Consumer Behaviour*, 14 (3), 193–207.

Molz, J. G. (2012) 'CouchSurfing and network hospitality: "It's not just about the furniture"', *Hospitality & Society*, 1 (3), 215-225.

Narasimhan, C., Papatla, P., Jiang, B., Kopalle, P. K., Messinger, P. R., Moorthy, S., … & Zhu, T. (2018) 'Sharing economy: Review of current research and future directions', *Customer Needs and Solutions*, 5 (1-2), 93-106.

OECD (2016) 'Des politiques propices à l'économie collaborative dans le tourisme', in *Éditions OCDE, Tendances et politiques du tourisme de l'OCDE*, Paris.

OECD (2020), 'Digital trends driving change in tourism', in *Éditions OCDE, Tourism Trends and Policies 2020*, Paris.

Parasuraman, A., Zeithaml, V. A. & Malhotra, A. (2005) 'ES-QUAL: A multiple-item scale for assessing electronic service quality', *Journal of Service Research*, 7 (3), 213-233.

Parasuraman, A., Zeithaml, V. A. & Berry, L. L. (1988) 'Servqual: A multiple-item scale for measuring consumer perceptions of service quality', *Journal of Retailing*, 64 (1), 12.

Perren, R. & Kozinets, R. V. (2018) 'Lateral exchange markets: How social platforms operate in a networked economy', *Journal of Marketing*, 82 (1), 20-36.

PwC (2016) 'Shared Benefits: How the Sharing Economy Is Reshaping Business across Europe', viewed 14 June 2020, http://www.pwc.co.uk/issues/megatrends/collisions/sharingeconomy/future-of-the-sharing-economy-in-europe-2016.html.

Sahlins, M. (1972) *Stone Age Economics*, New York: Aldine de Gruyter.

Sigala, M. (2017) 'Collaborative commerce in tourism: implications for research and industry', *Current Issues in Tourism*, 20 (4), 346-355.

Steylaerts, V. & Dubhghaill, S. O. (2012) 'CouchSurfing and authenticity: Notes towards an understanding of an emerging phenomenon', *Hospitality & Society*, 1 (3), 261-278.

Tussyadiah, I. P. & Pesonen, J. (2016) 'Impacts of peer-to-peer accommodation use on travel patterns', *Journal of Travel Research*, 55 (8), 1022-1040.

Tussyadiah, I. P. (2016) 'Factors of satisfaction and intention to use peer-to-peer accommodation', *International Journal of Hospitality Management*, 55, 70–80.

Wirtz, J., So, K., Mody, M., Liu, S. & Chun, H. (2019) 'Platforms in the peer-to-peer sharing economy', *Journal of Service Management*, 30 (4), 452-483.

Yang, S., Song, Y., Chen, S. & Xia, X. (2017) 'Why are customers loyal in sharing-economy services? A relational benefits perspective', *Journal of Services Marketing*, 31 (1), 48–62.

Part 3
Sustainable Tourism Development

8 Host–tourist interactions and residents' attitudes towards sustainable tourism development

Foad Irani, Ali Öztüren and Arash Akhshik

The objectives of this chapter are to:

- ☐ Investigate how the interaction between local community and visitors influences the residents' attitude.
- ☐ Assess the results of positive or negative attitudes on overall tourism development of an island.
- ☐ Suggest the establishment of roles and regulation for protecting the island resources.
- ☐ Suggest further research on sustainable tourism development in developing nations.

Keywords: host-tourist interactions, resident perceptions, resident attitudes, sustainable tourism, tourism development, North Cyprus.

Introduction

Tourism is often prescribed as a powerful catholicon for socio-economic development of island states (Chen et al., 2011; Ribeiro et al., 2013). Although tourism provides some benefits for residents and destinations, its introduction creates adverse impacts as a by-product of its functioning that erodes sustainable development goals (Alipour et al., 2020); therefore, in order to minimize the unintended impacts, the involvement of all stakeholders in the planning process is recommended by many scholars (e.g. Chang & Mak, 2018; Cusick, 2009; Ribeiro et al., 2018). Amongst the many stakeholders einvolved in the sustainable tourism management of an island, residents have been considered as the most important and effective, without whom any plans towards sustainable development goals may be doomed to failure (Lalicic & Önder, 2018).

Investigating the attitude of residents about the impacts of any development plan is gaining momentum in tourism, as an essential concern in managing and marketing sustainable tourism products and services (Chen & Chen, 2010; Choi & Murray, 2010; Kebete & Wondirad, 2019; Wang & Chen, 2015). Accordingly, the literature has been enriched with a variety of different studies related to residents' attitudes, interactions with tourists and different aspects of sustainable development, yet the ambiguity in the perception of these topics by locals is far from adequately explored (Eusebio et al., 2018). Therefore, it is imperative to delve into the host-tourist interactions to have a richer understanding of the residents' role in sustainable tourism development. Notably, it is vital to consider tourism as one of the biggest industries considered beneficial in developing countries and small state islands due to political and development nature of these areas, yet, there is a lack of adequate studies specifically in the developing nations (Akhshik et al., 2020; Sirivongs & Tsuchiya, 2012).

The attitude of the residents about sustainable tourism development is a concept that has received different definition by different scholars (Chen & Raab, 2012), but yet there is no clear-cut definition for the term (Ribeiro et al., 2013; Woosnam, 2012). Since there is no universally accepted definition, sustainable tourism development has been defined in a variety of explanations, and different theories have been employed (Wang & Chen, 2015). However, Sharpley (2014) categorizes the formation of these attitudes as intrinsic and extrinsic. On the other hand, the behavioral dimension of residents' attitudes was used in most of the significant studies of measures of support for sustainable tourism development (Woosnam, 2012). Thus, the same approach adopted in this study. One of the leading tourism experience components is the quality of the interactions between residents and tourists, that is capable of influencing both residents' and tourists' satisfaction either positively or negatively (Andereck et al., 2005; Kastenholz et al., 2018; Luo et al., 2015; Pizam et al., 2002).

There is a variety of conceptualizations of host-tourist interaction in the literature. A careful review of the related literature concludes that scholars should pay more attention to the sustainable aspect of marketing for more clarifications toward understanding the antecedents relevant to residents' attitudes (Eusebio & Carneiro, 2012; Sharpley, 2014). Although different factors may influence the residents' attitudes towards sustainable tourism development, there are different approaches to adopting the factors. However, host-tourist interactions have received less attention in the literature compared with other factors, such as perceptions of tourism impacts and community attachment (Eusebio et al., 2018). Consequently, there is a call to conduct more studies to explore the abovementioned factors that may influence tourists' attitudes toward sustainable tourism development in developing countries (Chen & Chen, 2010; Sharpley, 2014; Woosnam, 2012).

This empirical study, in North Cyprus, intended to examine the effects of three determinants, such as host-tourist interactions, and residents' attitude through positive and negative impacts of perception towards sustainable tourism development, by using a structured equation model (SEM). Therefore, the study will answer the question of whether the interactions between the local community and tourists influences the locals' attitudes towards sustainable development in the tourism industry. Thus, the objective of the study is narrowed down to: a) investigate how the interaction between local community and visitors influences the residents' attitudes; and b) assess the results of positive or negative attitudes on overall tourism development of an island. Following a literature review of relevant studies, a theoretical model has been developed and conceptually tested. The study also contributes to the theory and practice, the details of which is provided in the conclusion section.

Literature review

Scholars have contradictory definitions for residents' attitudes towards sustainable tourism development (e.g. Akis et al., 1996; Nunkoo & Gursoy, 2012; Ribeiro et al., 2013; Wang & Chen, 2015; Woosnam, 2012; Yeager et al., 2019). As a result, there is no clear-cut definition of this phenomena in the tourism in the literature (Chen & Raab, 2012; Latkova & Vogt, 2012;

Vargas-Sanchez et al., 2015). Nevertheless, the majority of literature leans towards sustainable tourism development as that which has a long-term concern for places, people, behaviors or other dimensions of the environment (Carmichael 2000; Gu & Ryan, 2008). Carmichael (2000) argued that residents' attitudes towards sustainable tourism development consist of three dimensions: cognitive, affective and behavioral. The cognitive dimension includes the beliefs, knowledge, and perception; the affective dimension can be described either as likes or dislikes; and the last dimension, behavioral, is explained as an expression of actions taken (Carmichael, 2000). The behavioral dimension was most frequently examined in tourism studies (Eusebio et al., 2018), and has been measured in the majority of studies through residents' participation (Choi & Murray, 2010; Gursoy & Rutherford, 2004; Latkova & Vogt, 2012), therefore in this study a similar approach has been adopted.

The formation of these attitudes is either through intrinsic or extrinsic factors (Sharpley, 2014). The literature reveals that there are different intrinsic and extrinsic factors influencing residents' attitude towards sustainable tourism development (Eusebio et al., 2018; Kastenholz et al., 2018). The intrinsic factors are the socio-demographic profile of residents, the perceived impacts of tourism, the benefits to individuals of the destinations' economy relying on the tourism industry, the attachment of the community to tourism activities, community involvement/participation, community satisfaction, place attachment, interaction with tourists, and residents' environmental attitudes; while the extrinsic factors are tourism development in the destination, the state of the local economy, and visitor types (Eusebio et al., 2018). However, a few factors have received less attention in tourism contexts, such as attachment to place or location, and interactions between the local community and visitors (Eusebio et al., 2018).

On the other hand, the impact of residents' attitudes towards sustainable tourism development has been rarely reported in less developed regions, specifically in island states. An exception is the study by Canalejo et al. (2015) that measures residents' attitudes in the Island of Santo Antao. Their study stresses the involvement of different stakeholders, specifically residents, as an essential element for economic and environmental sustainability in these specific geographical areas. Similarly, Sanchez Canizares et al. (2016) investigated tourists, residents, and business owners' attitudes as these are the pre-eminent stakeholders of the tourism industry. Moreover, the majority of studies allude to the positive attitude of residents' toward sustainable tourism activities (Lopez et al., 2018). However, the literature is still far from providing an adequate understanding of conceptualization in triggers influencing the residents' attitude.

Proposed model

The effects of variables on residents' attitude towards sustainable tourism development in North Cyprus was analyzed based on social exchange theory (SET) (Figure 8.1).

Figure 8.1: Proposed model of the study

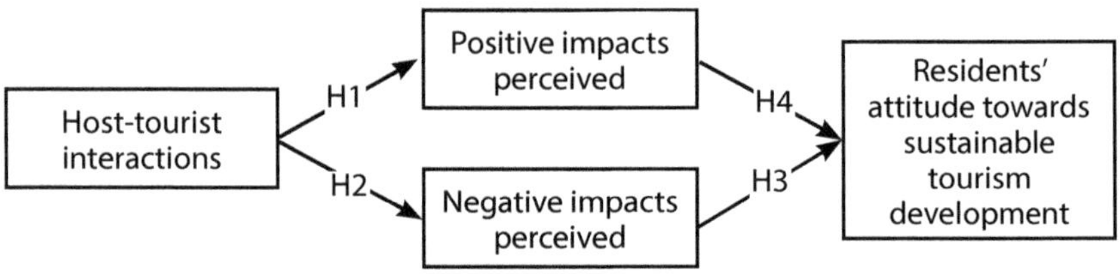

Source: Authors' elaboration.

Eusebio et al. (2018) emphasized in their study that host-tourist interaction is one of the determinants that significantly affects the residents' attitudes towards sustainable tourism development. The proposed model includes four variables. There are three determinants for residents' attitudes towards sustainable tourism development: host-tourist interactions; positive impacts perceived by residents; and negative impacts perceived by residents.

Interactions

The central components of the tourism experience are the confrontations between residents and tourists (Luo et al., 2015), that influence the perception of host and tourists positively or negatively (Andereck et al., 2005; Kastenholz et al., 2018; Pizam et al., 2002). Eusebio et al. (2018) argued that host-tourist interaction is a complicated construct, and has been studied by adopting different approaches. The interaction between host and tourist can be defined as social contact in tourism. In this study, we adopted Reisinger and Turner's (2003) perspective. Host-tourist interaction is an emerging construct, and due to the lack of adequate empirical studies there are ambiguities about the best method for measuring it (Eusebio et al., 2016; Kastenholz et al., 2013; Pizam et al., 2002; Reisinger & Turner, 2003).

This construct is used frequently in recent studies (Eusebio et al., 2016; Kastenholz et al., 2013) since the interaction between host and tourist was examined positively (Eusebio & Carneiro, 2012; Reisinger & Turner, 2003).

Eusebio et al. (2018) proved that in order to develop positive, sustainable tourism impacts, the interaction between host and tourist is a crucial factor, as the residents perceive more positive sustainable tourism impacts then they develop more positive attitude toward sustainable tourism. The characteristics and motivations of residents and tourists can be described in different varieties that influence the nature of the interaction between residents and tourists (Sharpley, 2014). The social contacts between residents and tourists almost are minimal and temporarily in the majority of the destinations (De Kadt, 1979; Eusebio & Carneiro, 2012; Kastenholz et al., 2015).

The interaction between residents and tourists can be either positive or negative due to the different characteristics of both actors (Luo et al., 2015; Reisinger & Turner, 2003). Cultural enrichment and learning about each other happens when both actors' perception are positive (Reisinger & Turner, 2003). A few studies investigated the relationship between host-tourist interaction and residents' attitudes towards sustainable tourism development (Eusebio & Carneiro, 2012; Eusebio et al., 2018; Kastenholz et al., 2013). Moreover, increasing the awareness among residents about sustainable tourism development and its advantages increases the positive attitude of residents towards tourism (Luo et al., 2015).Therefore, according to the literature, the following hypotheses developed:

- H1: Host-tourist interactions affect the positive impacts perceived.
- H2: Host-tourist interactions affect the negative impacts perceived.

Residents' perceptions

The residents' life quality can be influenced through economic, environmental, and socio-cultural determinants positively or negatively (Almeida-Garcıa et al., 2016; Latkova & Vogt, 2012). The history of tourism studies illustrates that although positive impacts of tourism were studied by researchers, the scholars in 1980s focused on both positive and negative impacts of sustainable tourism development (McGehee & Andereck, 2004). The literature review shows that residents perceive economic impacts as positive, meanwhile the other determinants such as environmental and socio-cultural impacts are perceived by residents as negative impacts of sustainable tourism development (Cheng et al., 2019; Gu & Ryan, 2008; Haralambopoulos & Pizam, 1996; Yoon et al., 2001). However, the perceptions of residents' influence their attitude both positively and negatively towards sustainable tourism development through economic, environmental, and socio-cultural determinants (Brida et al., 2011; Yoon et al., 2001).

In a considerable number of studies SET has been used to explore the effects of residents' perception on their attitude towards sustainable tourism development (Ali et al., 2018; Brida et al., 2011; Hadinejad et al., 2019; Lalicic & Önder, 2018; Pham et al., 2019; Rasoolimanesh et al., 2015). During the last decade, a considerable number of studies were pursued to explain the relationship between residents' perceptions and residents' attitudes towards sustainable tourism development (Diedrich & Garcia-Buades, 2009; McGehee & Andereck, 2004; Rasoolimanesh et al., 2019; Teye et al., 2002). SET is the dominate theory to explain the relationship between residents' perceptions and attitudes towards sustainable tourism development (Hsu et al., 2019). There is a variety of overarching theories to assess the residents' perception of sustainable tourism development activities.

The literature review revealed that SET is the prevailing discourse (Ap, 1990, 1992; Latkova & Vogt, 2012; Rasoolimanesh et al., 2017). Ap (1992) defined the SET that implements a cost-benefit analysis through the interaction of two parties. The interaction of host and tourist attitudes towards sustainable tourism development was conducted based on SET in different studies (Ali et al., 2018; Harrill, 2004; Kang & Lee, 2018; Long, 1996). Kang and Lee (2018) used SET to investigate the role of impacts on perception by the host community and their support for enhancing sustainable tourism development (Park et al., 2015; Rasoolimanesh et al., 2015; Yoon et al., 2001). Based on SET, residents support sustainable tourism development when they perceive the benefits; on the other hand, when residents perceived the cost they show less willingness to support sustainable tourism development (Chen & Chen, 2010; Kaltenborn et al., 2008; Lee, 2013; Meimand et al., 2017; Nunkoo & Gursoy, 2012).

Regarding the proposed model and according to the literature, the following hypotheses were developed:

- H3: Negative impacts perceived affects residents' attitudes.
- H4: Positive impacts perceived affects residents' attitudes.

Methodology

The majority of studies on the attitude of the local community towards sustainable tourism development adopted quantitative research method (Nunkoo et al., 2013; Sharpley, 2014). Recently, a considerable number of studies adopted structural equation modelling (SEM) to analyses the potential influence of residents' attitudes' antecedents towards sustainable tourism development (Chen & Chen, 2010; Dyer et al., 2007; Gu & Ryan, 2008;

Gursoy & Rutherford, 2004; Kaltenborn et al., 2008; Ko & Stewart, 2002; Lee, 2013; Lindberg & Johnson, 1997; Nunkoo & Ramkisoon, 2011; Nunkoo & Gursoy, 2012; Nunkoo et al., 2013; Oviedo-Garcia et al., 2008; Sirivongs & Tsuchiya, 2012; Vargas-Sanchez et al., 2011; Wang & Chen, 2015; Woosnam, 2012; Yoon et al., 2001). Eusebio et al. (2018) further investigate these attitudes using a single construct with a linear approach by an indirect measurement of impacts that forms the residents' attitudes towards sustainable tourism development.

A self-reported questionnaire was designed, based on the literature review to measure the dimensions of this study, naming, negative and positive impacts perceived, host-tourist interactions, and residents' attitude towards sustainable tourism development. Well-structured measures from various studies were the impetus in the design of the questionnaire. Following the instructions of Podsakoff et al. (2003), many procedural biases have been regarded. As such, the questionnaire consisted of a cover page explaining the aim of the study and the confidentiality of the collected data. The second part consisted of the behavioral dimension of the construct of attitudes of the residents towards sustainable tourism development. This 5-items construct was adopted from the study of McGehee and Andereck (2004) and further approved by Jurowski and Gursoy (2004).

The third part consists of the perceptions of 10 items for positive aspects and 10 items for negative aspects of perceived impacts by residents that have been adopted from Haralambopoulos and Pizam (1996) and further tested by Andereck and Nyaupane (2011). This part also measured the host–tourist interactions based on a 3-items construct that was initially adopted from Teye et al. (2002). The questionnaire continued with the demographic profile of the respondents, i.e. age, marital status, income, education, nationality. The questionnaire was translated into the Turkish language following the back-translation technique (Brislin, 1970). A pilot study of 10 residents approved the readability and clarity of the items in the questionnaire.

The data for the empirical study has been collected in the northern part of Cyprus; the residents are mainly Turks from Turkey and Turks originally from Cyprus, and the population is almost 300,000 people living in 3 main districts. The island attracts almost 1.5 million foreign tourists per year (North Cyprus hoteliers associations). Following judgmental sampling technique and targeting only native residents of the island, a paper-pen questionnaire was administrated during December 2019. The sample size of 380 (for a single proportion of 0.50, a confidence level of 95%, and a margin of error of 5%) deemed adequate to form the population of the study. In this regards, four hundred residents were invited to participate in the survey out of which 326 valid questionnaires were extracted.

The sample was almost evenly distributed amongst the genders (44.47% Males; 55.52% Females). Majority of the respondents were married (66.87%) and most of them were aged between 30-49 (37.11%), followed by the age group of 50-64 (35.88%), age group of over 65 (17.17%) and age group of 18-29 (9.81%). In terms of nationality, according to the aim of the research, only local native residents of Turks (11.34%) and Turkish Cypriots (88.65%) were included. Their education background varied from no schooling at all (13.8%) to graduate and higher degrees (6.74%) and trade/technical and vocational training made the majority of the sample (30.98%) whose income was from less than 500 USD (17.17%) to more than 3000 USD (4.90) with most earning between 500 and 1000 USD (64.72).

Data analysis

The data was assessed based on the theoretical underpinning. As a result, a two-step approach to structural equation model (SEM) has been applied in order to evaluate the measurement model and hypothesis testing (Anderson & Gerbing, 1988). In the first step, confirmatory factor analysis (CFA) was employed to weigh the dimensionality of constructs. This process was followed by discriminant validity, convergent validity, and a reliability test using Cronbach's alpha tested the internal consistency of the constructing variables using IBM SPSS V.20 and AMOS V.24. In the second step, an SEM was built, and a series of the model fit statistic was extracted in AMOS V.24, the results of which are presented below. Table 8.1. represents the results of CFA, mean, composite reliability (CR), average variance extracted (AVE). CR ranges from 0.881 to 0.925, whereas AVE ranging from 0.534 to 0.70 while all loadings are significant and more than 0.60 that indicates the discriminant and convergent validity of the constructs (Anderson & Gerbing, 1988; Fornell & Larcker, 1981).

In the second step, the structural model has been evaluated by fit indices such as χ^2: 835.636; df:293; χ^2/df: 2.852; CFI: 0.90; SRMR: 0.058; RMSEA: 0.075 that represent the acceptable interpretation of standard recommended cut-off criteria (Hu & Bentler, 1999). The result of SEM is presented in Table 8.2.

Hypotheses 1 and 2 that suggest host-tourist interactions have a positive effect on positive and negative impacts perceived by the residents are supported (H1: *t*-value: 8.031, $p < 0.001$; H2: *t*-value: 5.323, $p < 0.001$).

Table 8.1: Result of confirmatory factor analysis, CR, AVE and Alpha

	β	Mean	CR	AVE	α
Negative Impacts Perceived (Nip)		3.25	0.919	0.534	0.731
Increase of locals' stress	.654 *				
Increase of sexually transmitted diseases	.622 *				
Increase of road traffic	.761 *				
Decrease of peace and quiet	.709 *				
Increase in the goods & services' price level	.759 *				
Increase in criminality (robbery, violence)	.785 *				
Increase in real estate value	.802 *				
Increase of prostitution	.738 *				
Increase of drug consumption	.707 *				
Increase in vandalism acts	.754 *				
Positive Impacts Perceived (Pip)		3.88	0.925	0.608	0.780
Improvement of the island's image	.763 *				
Tourism contributes to the increase in residents' quality life	.756 *				
Infrastructure improvement (roads, sports facilities, etc.)	.843 *				
Increase in cultural events offering	.785 *				
Promotion of traditions	.804 *				
Growth of employment	.766 *				
Increase of the residents' income	.783 *				
Tourism improved public services in our community.	.732 *				
Attitudes towards sustainable tourism development		3.83	0.881	0.598	0.773
I would like to see more tourists on my island	.814 *				
I strongly support tourism development	.804 *				
I actively participate in tourism planning	.812 *				
I participate in the enhancement of tourism projects	.759 *				
I make suggestions for improvements in tourism development	.667 *				
Host-tourist Interactions (Int)		3.73	0.874	0.700	0.836
I like to interact with tourists	.914 *				
My interaction with tourists is positive	.819 *				
I make friends with tourists	.769 *				

Note: β: standardized factor loading; β is significant at the .001 level; AVE: average variance extracted; α: Cronbach's alpha; CR: composite reliability; *: $p < 0.001$

Source: Authors' elaboration.

Moreover, Hypotheses 3 and 4 that propose negative and positive impacts perceived by the residents have effects on residents' attitudes towards sustainable tourism development are supported (H3: *t*-value: 6.081, $p < 0.001$; H4: *t*-value: 9.065, $p < 0.001$).

Table 8.2: Result of the structural equation modeling (SEM)

Hypothesis	Estimate	S.E.	*t*-value	*p*	Result
Int → Nip	.453	.056	8.031	*	Supported
Int → Pip	.231	.043	5.323	*	Supported
Nip → Att	.296	.049	6.081	*	Supported
Pip → Att	.568	.063	9.065	*	Supported
Note: * $p < 0.001$					

Source: Authors' elaboration

Conclusion

The findings of this study provide essential theoretical and practical implications. This study has extended the theory of social exchange to the field of sustainable tourism development by giving a full consideration of host-tourist interactions on residents' attitude towards sustainable tourism development through positive and negative impacts perceived. The proposed model is unique and was developed based on SET. In this study, a proposed model developed and examined the relationship between residents' perception impacts both (positively and negatively) and the attitude of the local community towards sustainable tourism development in North Cyprus Island. As to the contributions to the literature, this study based on the previous studies, confirms the supportiveness of host-tourist interactions to the residents' attitudes towards sustainable tourism development based on SET (Eusebio et al., 2018; Hsu et al., 2019; Nunkoo et al., 2010; Ward & Berno, 2011). SET has been used frequently in the literature to explain the relationship between the residents' perception and residents' attitude towards sustainable tourism development. Four hypotheses developed according to the proposed model in this study. All four hypotheses were supported.

First, the variable of host-tourist interactions was found to have significant effects on both positive and negative impacts perceptions; thus, both H1 and H2 were supported. These findings are in line with prior studies (Eusebio et al., 2018). Furthermore, host-tourist interactions can be perceived either positively or negatively by residents, which means a good encounter between host and tourist leads to the willingness of residents to form positive attitudes towards sustainable tourism development. On the other hand, negative impacts perceived decrease the supportiveness of residents towards sustainable tourism development.

The other studies in the same context confirmed that host-tourist interactions significantly affect residents' attitudes towards sustainable tourism development through perceived impacts (Eusebio et al., 2018). Which means, the results of host-tourist encounters can lead to support for sustainable tourism development when there are perceived benefits, especially in island states because their economy relies on tourism. Of the two remaining hypotheses of the perceptions of positive and negative impacts on residents' attitudes towards sustainable tourism development, both H3 and H4 were statistically significant, thus were supported. These findings were based on SET: when residents perceive the benefits to be more than costs, their attitudes form positively to support for sustainable tourism activities, and on the contrary, when perceiving costs to be more than benefits, their attitudes form negatively towards sustainable tourism development. These findings confirmed previous studies (Eusebio et al., 2018).

The results of this study deliver prescriptions for tourism destination management on enhancing the involvement of residents in sustainable tourism development. There are several practical implications obtained from this study that explain the necessity of residents' participation in sustainable tourism development, particularly for island states. First, the government should support the non-governmental groups established by residents and involve the local community, under the supervision of government for sustainable tourism development. Furthermore, the government should provide a pathway to foster investment for any activities based on sustainable tourism development (Nunkoo et al., 2010). On the one hand, effective communication and interaction between residents and tourists at the destination is an essential antecedent of forming residents' attitudes. The local government, as a relevant organizer, should use its resources to establish proper interactions between residents and tourists to form positive and effective perceptions toward sustainable tourism development.

Second, the government should establish roles and regulation for protecting the island resources. Thus, beneficiary outcomes earned from tourism activities should be invested in development that results in more participation of the local community in sustainable tourism. The hospitality of residents significantly influences the success of sustainable tourism development (Bimonte & Punzo, 2016). Residents may provide supports toward tourists when they become aware of the financial and social benefits of sustainable tourism development (Lin et al., 2017). Therefore, the government has a crucial responsibility to shape positive attitudes by clarifying the importance of sustainable tourism development.

This study has some limitations. Tourism consist of a variety of stakeholders who are involved in the development of the sustainable tourism industry. In this study, the first limitation is that the questionnaires were gathered from local participants of North Cyprus. In order to significantly improve these kinds of studies in the same context, it is suggested to involve other stakeholders in examining the relationship between the perceptions and attitudes towards sustainable development of the tourism industry. Second, this study has been done in North Cyprus which has three main cities. To overcome this limitation, it is suggested that similar studies should be done on other islands. Finally, the data were gathered through the cross-sectional approach. It is recommended to collect data in future studies longitudinal to observe the differences that may take place.

References

Akhshik, A., Rezapouraghdam, H., & Ramkissoon, H. (2020). 'Industrialization of nature in the time of complexity unawareness: The case of Chitgar Lake, Iran'. *Journal of Hospitality & Tourism Research*, 1096348020964949.

Akis, S. Peristianis, N. & Warner, J. (1996) 'Residents' attitudes to tourism development: the case of Cyprus', *Tourism Managemen*t, 17 (7), 481-494.

Ali, L., Mi, J., Shah, M., Shah, S.J., Khan, S., Ullah, R. et al. (2018) 'Local residents' attitude towards road and transport infrastructure (a case of China Pakistan economic corridor)', *Journal of Chinese Economic and Foreign Trade Studies*. 11 (1), 104-120.

Alipour, H., Rezapouraghdam, H. & Akhshik, A. (2020) 'Redemption and the curse of tourism: The case of world's last inhabited troglodyte village' *Tourism and Planning & Development*, 1-18.

Almeida-García, F., Peláez-Fernández, M., Macias, R. & Balbuena, A. (2016) 'Residents' perceptions of tourism development in Benalmádena (Spain)', *Tourism Management*, 54, 259-274.

Andereck, K. L. & Nyaupane, G. P. (2011). 'Exploring the nature of tourism and quality of life perceptions among residents'. *Journal of Travel Research*, 50 (3), 248-260.

Andereck, K.L., Valentine, K.M., Knopf, R.R. & Vogt, C.A. (2005) 'Residents' perceptions of community tourism impacts', *Annals of Tourism Research*, 32 (4), 1056-1076.

Anderson, J. C. & Gerbing, D. W. (1988). 'Structural equation modeling in practice: A review and recommended two-step approach'. *Psychological Bulletin*, 103 (3), 411.

Ap, J. (1990) 'Residents' perceptions research on the social impacts of tourism', *Annals of Tourism Research*, 17 (4), 10-616.

Ap, J. (1992) 'Residents' perceptions on tourism impacts', *Annals of Tourism Research*, 19 (4), 665-690.

Bimonte, S. & Punzo, L.F. (2016) 'Tourist development and host–guest interaction: An economic exchange theory', *Annals of Tourism Research*, 58, 128-139.

Brida, J.G., Osti, L. & Faccioli, M. (2011) 'Residents' perception and attitudes towards tourism impacts: A case study of the small rural community of Folgaria (Trentino-Italy)', *Benchmarking*, 18 (3), 359-385.

Brislin, R. W. (1970). 'Back-translation for cross-cultural research'. *Journal of Cross-Cultural Psychology*, 1 (3), 185-216.

Canalejo, A., Núñez-Tabales, J., López, J. & Fuentes García, F. (2015) 'Resident perceptions of community tourism in Cape Verde', *Tourism Culture & Communication*, 15 (2), 103-119.

Carmichael, B.A. (2000) 'A matrix model for resident attitudes and behaviours in a rapidly changing tourist area', *Tourism Management*, 21 (6), 601-611.

Chang, R.C. & Mak, A.H. (2018) 'Understanding gastronomic image from tourists' perspective: A repertory grid approach', *Tourism Management*, 68, 89-100.

Chen, C.F. & Chen, P.C. (2010) 'Resident attitudes toward heritage tourism development', *Tourism Geographies*, 12 (4), 525-545.

Chen, C.M. Chen, S.H. & Lee, H.T. (2011) 'The destination competitiveness of Kinmen's tourism industry: exploring the interrelationships between tourist perceptions, service performance, customer satisfaction and sustainable tourism', *Journal of Sustainable Tourism*, 19 (2), 247-264.

Chen, S. & Raab, C. (2012) 'Predicting resident intentions to support community tourism: Toward an integration of two theories', *Journal of Hospitality Marketing & Management*, 21 (3), 270-294.

Cheng, T.M., Wu,H.C., Wang, J.T-M. & Wu, M-R. (2019) 'Community participation as a mediating factor on residents' attitudes towards sustainable tourism development and their personal environmentally responsible behaviour', *Current Issues in Tourism*, 22 (14), 1764-1782.

Choi, H.C. & Murray, I. (2010) 'Resident attitudes toward sustainable community tourism', *Journal of Sustainable Tourism*, 18 (4), 575-594.

Cusick, J. (2009) 'At the intersection of resident, research andrecreation stakeholder interests: East Maui, Hawai'i, as a sustainable tourism destination', *Island Studies Journal*, 4 (2).

De Kadt, E. (1979) 'Social planning for tourism in the developing countries', *Annals of Tourism Research*, 6 (1), 36-48.

Diedrich, A. & García-Buades, E. (2009) 'Local perceptions of tourism as indicators of destination decline', *Tourism Management*, 30 (4), 512-521.

Dyer, P., Gursoy,D., Sharma,B. & Carter,J. (2007) 'Structural modeling of resident perceptions of tourism and associated development on the Sunshine Coast, Australia', *Tourism Management*, 28 (2), 409-422.

Eusebio, C. & Carneiro, M.J. (2012) 'Socio-cultural impacts of tourism in urban destinations', *Revista Portuguesa de Estudos Regionais*, 30 (1), 65-76.

Eusebio, C. Carneiro, M.J. & Caldeira, A. (2016) 'A structural equation model of tourism activities, social interaction and the impact of tourism on youth tourists' QOL', *International Journal of Tourism Policy*, 6 (2), 85-108.

Eusebio, C. Vieira, A.L. & Lima, S. (2018) 'Place attachment, host–tourist interactions, and residents' attitudes towards tourism development: The case of Boa Vista Island in Cape Verde', *Journal of Sustainable Tourism*, 26 (6), 890-909.

Fornell, C. & Larcker, D. F. (1981). 'Structural equation models with unobservable variables and measurement error: Algebra and statistics'. *Journal of Marketing Research*, 18, 382-388.

Gu, H. & Ryan, C. (2008) 'Place attachment, identity and community impacts of tourism—the case of a Beijing huton'g, *Tourism Management*, 29 (4), 637-647.

Gursoy, D. & Rutherford, D.G. (2004) 'Host attitudes toward tourism: An improved structural model', *Annals of tourism Research*, 31 (3), 495-516.

Hadinejad, A., Nunkoo, R., Moyle, B., Scott, N. & Kralj, A (2019) 'Residents' attitudes to tourism: a review', *Tourism Review*. 74. 10.1108/TR-01-2018-0003.

Haralambopoulos, N. & Pizam, A. (1996). 'Perceived impacts of tourism: The case of Samos'. *Annals of Tourism Research,* 23 (3), 503-526.

Harrill, R. (2004) 'Residents' attitudes toward tourism development: A literature review with implications for tourism planning'. *Journal of Planning Literature,* 18 (3), 251-266.

Hsu, C.Y. Chen, M.Y. & Yang, S.C. (2019) 'Residents' attitudes toward support for island sustainable tourism', *Sustainability,* 11 (18), 5051.

Hu, L. T. & Bentler, P. M. (1999). 'Cut-off criteria for fit indexes in covariance structure analysis: Conventional criteria versus new alternatives'. *Structural Equation Modeling,* 6 (1), 1-55.

Jurowski, C., & Gursoy, D. (2004). 'Distance effects on residents'attitudes toward tourism'. *Annals of Tourism Research,* 31 (2), 296-312.

Kaltenborn, B.R.P., Andersen,O., Nellemann,C., Bjerke, T. & Thrane, C. (2008) 'Resident attitudes towards mountain second-home tourism development in Norway: The effects of environmental attitudes', *Journal of Sustainable Tourism,* 16 (6), 664-680.

Kang, S.K. & Lee, J. (2018) 'Support of marijuana tourism in Colorado: A residents' perspective using social exchange theory', *Journal of Destination Marketing & Management,* 9, 310-319.

Kastenholz, E. Carneiro, M.J. & Eusebio, C. (2018) 'Diverse socializing patterns in rural tourist experiences–a segmentation analysis', *Current Issues in Tourism,* 21 (4), 401-421.

Kastenholz, E., Carneiro, M.J., Eusébio, C., & Figueiredo, E. (2013) 'Host–guest relationships in rural tourism: Evidence from two Portuguese villages', *Anatolia,* 24 (3), 367-380.

Kastenholz, E., Eusebio, C. & Figueiredo, E. (2015) 'Contributions of tourism to social inclusion of persons with disability', *Disability & Society,* 30 (8), 1259-1281.

Kebete, Y. & Wondirad, A. (2019) 'Visitor management and sustainable destination management nexus in Zegie Peninsula, Northern Ethiopia', *Journal of Destination Marketing & Management,* 13, 83-98.

Ko, D.W. & Stewart, W.P. (2002) 'A structural equation model of residents' attitudes for tourism development', *Tourism Management,* 23 (5), 521-530.

Lalicic, L. & Önder, I. (2018) 'Residents' involvement in urban tourism planning: Opportunities from a smart city perspective', *Sustainability,* 10 (6), 1852.

Látková, P. & Vogt, C.A. (2012) 'Residents' attitudes toward existing and future tourism development in rural communities', *Journal of Travel Research*, 51 (1), 50-67.

Lee, T.H. (2013) 'Influence analysis of community resident support for sustainable tourism development', *Tourism Management*, 34, 37-46.

Lin, Z., Chen, Y. & Filieri, R. (2017) 'Resident-tourist value co-creation: The role of residents' perceived tourism impacts and life satisfaction', *Tourism Management*, 61, 436-442.

Lindberg, K. & Johnson, R.L. (1997) 'Modeling resident attitudes toward tourism', *Annals of Tourism Research*, 24 (2), 402-424.

Long, M. (1996) 'The role of the linguistic environment in second language acquisition', In W. C. Ritchie, & T. K. Bhatia (Eds.), *Handbook of Second Language \Acquisition* (pp. 413-468). New York: Academic Press.

López, M.F.B., Francisca,M., Virto,N.R., Manzano, J.A. & Miranda,J.G-M. (2018) 'Residents' attitude as determinant of tourism sustainability: The case of Trujillo', *Journal of Hospitality and Tourism Management*, 35, 36-45.

Luo, X. Brown, G. & Huang, S.S. (2015) 'Host perceptions of backpackers: Examining the influence of intergroup contact', *Tourism Management*, 50, 292-305.

McGehee, N. G. & Andereck, K. L. (2004). 'Factors predicting rural residents' support of tourism'. *Journal of Travel Research*, 43 (2), 131-140.

Meimand, S.E., Khalifah, Z., Zavadskas, E., Mardani, A., Najafipour, A. & Ahmad, U. (2017) 'Residents' attitude toward tourism development: A sociocultural perspective', *Sustainability*, 9 (7), 1170.

Nunkoo, R. & Gursoy, D. (2012) 'Residents' support for tourism: An identity perspective', *Annals of Tourism Research*, 39 (1), 243-268.

Nunkoo, R. & Ramkissoon, H. (2011) 'Developing a community support model for tourism', *Annals of Tourism Research*, 38 (3), 964-988.

Nunkoo, R. Gursoy, D. & Juwaheer, T.D. (2010) 'Island residents' identities and their support for tourism: an integration of two theories', *Journal of Sustainable Tourism*, 18 (5), 675-693.

Nunkoo, R., Ramkissoon, H. & Gursoy, D. (2013) 'Use of structural equation modeling in tourism research: Past, present, and future', *Journal of Travel Research*, 52 (6), 759-771.

Oviedo-Garcia, M.A. Castellanos-Verdugo, M. & Martin-Ruiz, D. (2008) 'Gaining residents' support for tourism and planning', I*nternational Journal of Tourism Research*, 10 (2), 95-109.

Park, D.B. Nunkoo, R. & Yoon, Y.S. (2015) 'Rural residents' attitudes to tourism and the moderating effects of social capital', *Tourism Geographies*, 17 (1), 112-133.

Pham, K. Andereck, K. & Vogt, C. (2019) 'Local residents' perceptions about tourism development'. *Travel and Tourism Research Association: Advancing Tourism Research Globally*, 74.

Pizam, A. Fleischer, A. & Mansfeld, Y. (2002) 'Tourism and social change: The case of Israeli ecotourists visiting Jordan', *Journal of Travel Research*, 41 (2), 177-184.

Podsakoff, P. M., MacKenzie,S.B., Lee, J-Y. & Podsakoff, N.P. (2003) 'Common method biases in behavioral research: a critical review of the literature and recommended remedies', *Journal of Applied Psychology*, 88 (5), 879.

Rasoolimanesh, S.M., Jaafara, M., Kockb, N. & Ramayahc, T. (2015) 'A revised framework of social exchange theory to investigate the factors influencing residents' perceptions', *Tourism Management Perspectives*, 16, 335-345.

Rasoolimanesh, S.M., Roldán, J., Jaafar, M. & Ramayah, T. (2017) 'Factors influencing residents' perceptions toward tourism development: Differences across rural and urban world heritage sites', *Journal of Travel Research*, 56 (6), 760-775.

Rasoolimanesh, S.M., Taheri,B., Gannon, M.,Ali Vafaei-Zadeh, A. & Hanifah, H. (2019) 'Does living in the vicinity of heritage tourism sites influence residents' perceptions and attitudes?, *Journal of Sustainable Tourism*, 27 (9), 1295-1317.

Reisinger, Y. & Turner, L. (2003) *Cross-Cultural Behaviour in Tourism*, Routledge.

Ribeiro, M.A. Valle, P.O.D. & Silva, J.A. (2013) 'Residents' attitudes towards tourism development in Cape Verde Islands', *Tourism Geographies*, 15 (4), 654-679.

Ribeiro, M.A., Pinto, P. & Silva, J. (2018) 'Examining the predictive validity of SUS-TAS with maximum parsimony in developing island countries', *Journal of Sustainable Tourism*, 26 (3), 379-398.

Sánchez Cañizares, S.M., Castillo Canalejo, A.M. & Núñez Tabales, J.M. (2016) 'Stakeholders' perceptions of tourism development in Cape Verde, Africa', *Current Issues in Tourism*, 19 (10), 966-980.

Sharpley, R. (2014) 'Host perceptions of tourism: A review of the research', *Tourism Management*, 42, 37-49.

Sirivongs, K. & Tsuchiya, T. (2012) 'Relationship between local residents' perceptions, attitudes and participation towards national protected areas: A case study of Phou Khao Khouay National Protected Area, central Lao PDR', *Forest Policy and Economics*, 21, 92-100.

Teye, V. Sirakaya, E. & Sönmez, S.F. (2002) 'Residents' attitudes toward tourism development', *Annals of Tourism Research*, 29 (3), 668-688.

Vargas-Sanchez, A., Porras-Bueno, N. & Plaza-Mejía, M. (2011) 'Explaining residents' attitudes to tourism: Is a universal model possible?', *Annals of Tourism Research*, 38 (2), 460-480.

Vargas-Sánchez, A., Pinto, P., Mendes, J. & Silva, J. (2015) 'Residents' attitude and level of destination development: An international comparison', *Tourism Management*, 48, 199-210.

Wang, S. & Chen, J.S. (2015) 'The influence of place identity on perceived tourism impacts', *Annals of Tourism Research*, 52, 16-28.

Ward, C. & Berno, T. (2011) 'Beyond social exchange theory: Attitudes toward tourists', *Annals of Tourism Research*, 38 (4), 1556-1569.

Woosnam, K.M. (2012) 'Using emotional solidarity to explain residents' attitudes about tourism and tourism development', *Journal of Travel Research*, 51 (3), 315-327.

Yeager, E.P., Boley, B.B., Woosnam, K.M. & Green, G.T. (2019) 'Modeling residents' attitudes toward short-term vacation rentals', *Journal of Travel Research*, 0047287519870255.

Yoon, Y., Gursoy, D. & Chen, J.S. (2001) 'Validating a tourism development theory with structural equation modeling', *Tourism Management*, 22 (4), 363-372.

9 Challenges to sustainability in prospective world heritage sites

Sina Kuzuoglu and Stella Kladou

The objectives of this chapter are to:

- ☐ Investigate how a centralized governance structure affects the potential to achieve sustainable tourism development.
- ☐ Explore whether the sustainability challenges that arise from the top-down management approach in heritage sites in Turkey proposed in 1990s improved over the years.
- ☐ Assess empirical data collected through interviews and focus groups in Iznik/Nicaea and use the sustainability framework as a basis for a thematic analysis on the impact of Iznik's prospective inscription as a World Heritage Site.
- ☐ Describe the role of coordination between stakeholder groups, previous urban development patterns, conservation of heritage assets and the natural landscape for potential sustainable tourism development.
- ☐ Address theoretical and practical implications of inscription as a World Heritage Site and tourism development on sustainability.

Keywords: cultural heritage; World Heritage Sites; sustainability; urban development; spatial planning.

Introduction

Heritage cities often build on their cultural assets to develop as destinations. Yet, the assets communicated to visitors usually present only part of the picture. Heritage cities often tend to possess a significant cultural value that is hidden from the prying eyes. This value is not only exhibited as tangible assets, but the accumulation of knowledge and experience in the culture-laden milieu is also central to the notion of heritage (Falser, 2015). Such intangible elements have found their way into international discussions in recent years (Vecco, 2010). Heritage cities, as living spaces, also embrace a cultural heritage inseparable from local communities, and in a broader perspective, from the whole of humanity. This perspective is central to the creation of World Heritage Sites (WHSs) seeking to protect the world's heritage with Outstanding Universal Value under the umbrella of the United Nations Educational, Scientific and Cultural Organization (UNESCO) (1972). In this process, however, the heritage assets inscribed as WHSs are nominated by national governments which, upon inscription, become the primary entity responsible for safeguarding their integrity.

The primary motivation for this research is the centralized governance structure in developing countries and how this affects the potential to achieve sustainability through tourism development. Turkey is an example of such a centralized system in regard to tourism and heritage management (Yüksel et al., 2005) which includes the WHSs. In Turkey, WHSs are not only effective tools to aid foreign policy (Atakuman, 2010) but are also integral to culture-driven tourism strategy (Ozturk & van Niekerk, 2014). Despite inconclusive evidence in the literature on the relationship between proliferation of tourism and WHS status (Jimura, 2011; Poria et al., 2013), Turkey's proactive relationship with UNESCO may be interpreted as re-affirming the perceived linkage between WHS status and visitation.

Furthermore, extant studies in Turkey argue that the central governance structure in Turkey commonly excludes local stakeholders from planning processes (Tosun & Jenkins, 1996; Yüksel et al., 1999). Still, in the subsequent years, scholarly interest in the governance of tourism and heritage in Turkey declined considerably and does not reveal if earlier recommendations have been taken into consideration. Although efforts are made to decentralize development mechanisms in Turkey and to increase region-based decision-making (Alvarez et al., 2014), their ramifications for tourism and heritage are virtually non-existent in the literature. This research aims to have a glimpse of the Turkish perspectives on heritage governance and to investigate whether the sustainability challenges that arise from the top-down management approach in heritage sites proposed in 1990s (Tosun, 1998) improved over the subsequent decades. To this end, this research places the ancient city of Iznik/Niceae in Bursa province of northwestern Turkey under the microscope.

Iznik, as a living city for the last three millennia, represents a unique accumulation of cultural heritage assets of different empires (Stefanidou, 2003). Despite being inhabited for its entire history, the city has had a tumultuous past, with shifting geo-political and economic importance (Lowry, 2003). Iznik's cultural importance for the national government resonates in the city's recent nomination as a prospective WHS (UNESCO, 2014). Many heritage cities are also living cities, thereby implying a series of challenges when WHS inscription, urban and tourism development and sustainability objectives are pursued. The literature tends to prioritize either heritage/conservation or destination development, with implications often favoring a centralized system in order to facilitate decision-making. The question arising, however, is whether those in charge do learn over time, grasp the essence of good/bad practice and are, in the end, in position to support sustainability. Since WHS status is an important asset for destination promotion of rural areas in particular (Wuepper & Patry, 2016), development objectives may prioritize economic growth and, in the process, neglect environmental and socio-cultural attributes.

Drawing on previous scholarship on WHSs and destination management, this research investigates Iznik's administrative framework pertinent to its tourism prospects and to-be WHS status for an ex ante evaluation of the sustainability challenges. To achieve this, we particularly focus on how the culture embedded in Iznik and its inhabitants is reflected in administrative stakeholders and how this reflection is interlinked with sustainability objectives in this heritage city.

Literature review

WHSs and heritage cities (e.g., those cities in which sites of the UNESCO World Heritage List are located) need to embrace future generations while extending beyond today's local communities to international visitors. As such, their sustainable development relies on three pillars, namely the economic, the socio-cultural and the environmental ones (United Nations World Tourism Organization [UNWTO], 2013). Yet, literature suggests that the policy, tourism and development objectives which are linked to each one of these pillars may not necessarily be reflected in practice when a living city is inscribed as a WHS by UNESCO.

Conservation in heritage cities: Friend or foe?

It is not an easy task to help heritage cities deal with the challenges of protecting and managing their heritage, while developing as destinations. The starting point in this challenge is the definition and recognition of heritage itself. Falser (2015:2) sees cultural heritage as "material structures, institutional complexes and practices, ... [that] carr[y] a powerful emotional charge and a value structure emanating from the idea of belonging and of shared cultural meanings", thereby embodying both the tangible and the intangible assets of a particular place under the umbrella of cultural heritage. However, the integration of intangible heritage into the international agenda is a relatively recent phenomenon (Vecco, 2010). Heritage can be appropriated in a bottom-up manner by the residents of a particular place or it may be designated in a top-down approach (Rautenberg, 1998 cited from Dupagne et al., 2004; Tweed & Sutherland, 2007) primarily emanating from a certain group of experts.

However, this approach often undervalues, and subsequently disregards, the public opinion in living cities in an attempt to regenerate the economic structure. This approach may lead to deterioration of the residents' quality of life and to negative impacts on the resident-visitor co-creation of value (Kavaratzis, 2017). An example of this is the City of Pingyao, China, whose designation as a WHS has led to severely negative socio-cultural impacts, such as gentrification and displacement (Wang, 2012). Aside from its societal consequences, a conservation framework may inflict an additional monetary pressure on residents, which is amplified through inscription as a WHS due to the international pressure for conservation (Joy, 2016). This possibility also aligns with the view that WHS management attaches a paramount importance to environmental management but fails to consider social parameters to the same extent (Landorf, 2009).

WHSs are essentially regarded as being the joint heritage of humanity, whose conservation is desired for the benefit of future generations (UNESCO, 1972). Yet, the conservation framework imposed by UNESCO for WHSs may not be aligned with the needs of the communities inhabiting the WHS. Especially for WHSs located in cities, the conservation expectations may not capture the city's developmental requirements (Pendlebury et al., 2009). Furthermore, in an era of global mobility, WHSs and other heritage assets are not solely connected to the local community. Rather, they are valorized by visitors as well, and the tourism activity they generate is perceived to positively influence the conservation efforts in heritage settings (Vareiro et al., 2013).

On the other hand, intangible elements of cultural heritage may be more difficult to preserve, particularly when one considers the intercultural milieu found in practically all tourism destinations (Andereck et al., 2005). Together with the physical attributes, the immaterial knowledge and experience that accumulate in cities interactively shape residents' attachment to their cities, i.e., their sense of place (Campelo et al., 2013). As such changes may influence a city's social fabric, a coherent and comprehensive planning process is essential to maintain attractiveness.

Tourism planning and development in heritage cities

Becoming a WHS does not necessarily translate to increased visitation (Jimura, 2011; Poria et al., 2013) and the negative examples of over-tourism in many prominent heritage cities (e.g. Venice), highlight potential outcomes of tourism mismanagement that both endanger the integrity of the heritage assets and the quality-of-life of the residents (Seraphin et al., 2018). Destinations, whether they be cities or not, are not static entities, their innate characteristics are transformed by external contingencies. Tourism may be regarded as one such contingency, continuously evolving in itself and influencing the destination characteristics (Buhalis, 2000; Butler, 1980). In heritage cities, as argued by Harrill and Potts (2003), tourism development may influence residents to different extents based on their proximity to core heritage areas.

Previous scholarship on the role of WHS status in destination management argues that rural areas integrate inscription into their marketing efforts to a higher extent in an attempt to decrease the expected dissatisfaction that may be expressed by prospective visitors prior to traveling to a relatively remote destination (Wuepper and Patry, 2016). When we approach this from a destination competitiveness framework (Buhalis, 2000), proliferation of tourism requires not only infra- and superstructure elements but also draws strength from experiential elements and peripheral services, which often are not as developed in rural areas compared to their urban counterparts. In living cities, however, these elements are not solely used by visitors but they also cater to the needs of the local community members, implying that tourism planning, ideally, incorporates the residents' needs into the planning process (Ashworth and Page, 2011) which renders the local and national tourism governance an integral component of sustainable tourism perspective of heritage cities.

Extant studies show that socio-cultural and economic characteristics of cities change with increasing tourism activity (e.g., Andereck et al., 2005;

Buhalis, 2000; Özel and Kozak, 2017; Vareiro et al., 2013). In Turkey's context, the centralized governance approach to tourism and heritage conservation is an important parameter that not only influences the planning phase (Human, 2015) but also shapes the execution stage (Yüksel et al., 2005).

Governance mechanisms and heritage sites in Turkey

Turkey exhibits a central governance structure in which national institutions are the primary decision-makers and local stakeholders are commonly excluded from planning processes (Human, 2015; Yüksel et al., 1999; Yüksel et al., 2005). While the recent creation of regional development agencies (RDAs) is an important step towards decentralization and addressing region-specific development needs (Alvarez et al., 2014), the progress of this process and its efficacy remains enigmatic. In the case of WHSs in Turkey, their selection and inscription processes are undertaken by the national government, in line with Rautenberg's (1998) heritage by designation approach, which appoints site managers to each nominated or inscribed WHS.

The general tendency of WHS management in Turkey involves local administrations to employ the experts to work on heritage assets within their legal jurisdiction (Smith, 2006) and provide financial support to the conservation and renovation efforts. However, different national entities are responsible for different resources, and, in Turkey, the absence of a coordination mechanism between different governmental branches and their local representatives presents a challenge for heritage cities, particularly in cities where natural and cultural heritage co-exist (for an example from Cappadocia WHS in Turkey, see Somuncu & Yiğit, 2007).

In the context of the present research, there are three primary administrative institutions. First, the Ministry of Tourism and Culture oversees the heritage aspect due to the potential WHS status of Iznik which, in addition to cultural heritage assets, draws strength from the residents' 'connect[ion] with the lake and their agricultural hinterland' (UNESCO, 2014). Second, the appointed site manager is responsible for the preparation of a management plan and the administration of the Field Management Unit (FMU) which ideally informs different governmental entities and represents the bridge between local and national stakeholders. Third, Iznik's FMU is staffed and funded by Bursa Metropolitan Municipality (BMM), which hierarchically supersedes the local Iznik Municipality (IM). Both municipalities are responsible for regional and local planning initiatives and thus, integral to heritage management and tourism planning in Iznik, and their ideal integration with sustainability objectives. Prior to turning attention to the methodological approach employed, a brief introduction on Iznik as a prospective WHS and its historical evolution is essential to provide a context for the present research.

The case of Iznik/Nicaea, Turkey

Iznik, or as was known in the olden times, Nicaea, is home to extensive archaeological studies and renovations undertaken with the ultimate aim of becoming a WHS, a process which started with the city's introduction to the tentative list in 2014 (UNESCO, 2014). In essence, Iznik is one of the few heritage cities in Turkey that openly exhibits a multilayered structure, in which different cultural influences have culminated in a unique representation of historical transformation.

Iznik is a district of Bursa, located in the north-western Marmara region of Turkey. First established in the 3rd century BC (Stefanidou, 2003), it was successively ruled by Bithynia, Roman, Seljuk, Byzantine and Ottoman Empires. Originally a walled city (i.e., fortress) established on the eastern part of its namesake lake and surrounded by mountains, Iznik exhibits a Hippodamian city plan and a grid layout that is primarily intact to date. Over the course of history, Iznik has become an important city for Christianity for hosting two ecumenical councils (UNESCO, 2014). It is also the city where the higher education system of the Ottoman Empire was shaped, and the first madrasah founded (Bayraktar, 1994). Furthermore, from 16th century onwards, Iznik became an important supplier of ceramics (i.e., çini) for the Ottoman Imperial Palace, which cooperated with the manufacturers on the ornamental style with sketches mostly completed in the Imperial Palace (Aslanapa, 2003).

While Iznik has been a continuously inhabited city since its inception, Lowry (2003) notes that it was largely underpopulated during the Ottoman Empire, especially after the relocation of the Ottoman capital to Constantinople (modern day Istanbul). According to the travelogue of Evliya Celebi, the eastern part of the city was mostly abandoned and in ruins in the 17th century (Kahraman & Dağlı, 2013), and its local economic and physical conditions further deteriorated after the 18th century (Foss, 2003; Lowry, 2003).

In line with the scope of the present research, there are three facts that would help us draw the picture of today's Iznik. First, the tangible cultural heritage assets in Iznik represent different cultures that have reigned over the city throughout the last millennia. However, their condition ranges from well-preserved to derelict. Second, Iznik has experienced a surge in its population after the demise of the Ottoman Empire in 1923. Third, not only did the section within the fortress develop significantly (which is not always aligned with the current conservation framework) but Iznik also expanded towards its periphery, implying a change in its land-use framework and a decline in the agricultural production (Yücel & Salt, 2018). Integration of a sustainability paradigm in a culturally endowed living city is especially difficult to achieve, due to innumerable interested stakeholder groups, which renders the methodological approach particularly important.

Methodology

A thematic analysis of secondary and primary data was deemed necessary. As a basis for the thematic analysis undertaken, we refer to the three pillars of sustainability, namely socio-cultural, economic and environmental, as these embrace objectives relevant to, among others, economic viability, local prosperity, social equity, visitor fulfillment, local control, community well-being, cultural richness, physical integrity, and resource efficiency (UNWTO, 2013).

Extant information on Iznik, including the output of international symposia on Iznik (Akbaygil et al., 2003; BMM, 2015; IM, 2005) and the nomination file of Iznik to the World Heritage Centre (UNESCO, 2014), reveal that cultural heritage assets constitute the overwhelming majority of scholarly interest in Iznik. The nomination file of Iznik summarizes its heritage identity in three main constituents, namely, tangible cultural heritage assets (e.g., the archaeological remains and architectural assets); intangible cultural heritage assets, (e.g., çini art, its importance for Christianity); and the continuous co-existence of nature and humans for millennia (UNESCO, 2014). Any reference to the urban development around Iznik is virtually non-existent. However, the cultural landscape changed considerably due to modifications in land-use plans (Yücel & Salt, 2018).Therefore, we reviewed the urban development plan at 1/1000 implemented in line with 1/5000 strategic development plans, and the recent revisions made on them.

Primary data collection followed a qualitative approach, which included a combination of observations, focus groups, and interviews. As the UNESCO procedure is at its infancy, the people knowledgeable on the current status of Iznik are limited. At first, the local UNESCO project team facilitated an in-depth briefing on the history and cultural heritage assets of Iznik and its UNESCO process. Then, given that our research requires site-specific information as well as more general knowledge on Iznik, two research visits to Iznik were arranged in spring 2018. During these visits, observational data were collected, one interview and two focus groups were held. The topics were determined a priori on the basis of the literature review according to the participants' professional backgrounds. Upon processing these data, additional topics emerged; these were discussed during an additional interview with the Iznik Field Manager.

In detail, our first research trip in March 2018 primarily focused on a preliminary assessment of the cultural heritage assets in Iznik. During this trip, we collected observational information on the city and its tourism assets. Furthermore, we interviewed one of the most prominent çini masters of Iznik according to the BMM representatives in order to determine the characteristics of çini, important actors in the industry, threats to its continuity, and its limitations.

Our second research trip in April 2018 was guided by representatives of the BMM. Iznik can be regarded as an example of heritage by designation (Rautenberg, 1998), and the team working in the city are experts in different disciplines of heritage assets. Our guides and accompanying colleagues (i.e., two archaeologists, two art historians, a UNESCO field expert and an NGO representative) were the participants of our first focus group, in which we acquired information both on cultural heritage assets in Iznik, the history of Iznik and also on different aspects of tourism activity in Iznik. In the second focus group, we met with the administrators of IM and the Public Education Center (PEC), which focused on the identity of and human capital in Iznik, and the administrative perspective surrounding tourism development and conservation efforts. In brief, our informants represent higher sections of the bureaucratic ladder that works in Iznik. The information provided by our respondents and interviewers were continuously compared to the existing body of literature on Iznik (Glaser & Strauss, 1967). Since the UNESCO process is ongoing and relevant authorities did not wish to engage with the general public at the initial stage, we could not engage with local community members during this exploratory study. Thus, top-down stakeholders were the ones to respond to our inquiries on the upcoming residents' involvement in decision-making process.

Finally, in September 2018, a later interview with the Iznik Field Manager evolved around the projected timeline of the UNESCO procedure and the possible spatial interventions that would aid the inscription process of Iznik and in its aftermath. During this interview, in an attempt to evaluate managerial initiatives from a sustainability lens, we further focused on whether and how the FMU indented to embrace residents in decision-making.

Data collection was conducted in Turkish and the focus group meetings and the interviews lasted 35-50 minutes each, were recorded and transcribed. The transcriptions were subsequently coded with the primary aim of determining recurring information and themes, as well as to discover differences in the participants' respondents (Ryan & Bernard, 2003). Data collection was conducted in Turkish and then translated by the authors.

Results and discussion

Drawing on the background provided for Iznik, there are numerous resources in Iznik that may be integrated into the city's tourism offering portfolio. However, in line with our research focus on the sustainability aspect of Iznik's future tourism prospects, interest revolves mainly around socio-cultural and economic sustainability, since these emerge as the priorities for the inscription of the city. However, the study seeks to include some reference to the environmental outcomes of tourism development, and how this interrelates with the conservation framework in Iznik that is essential for its inscription. Extant studies in the Turkish context highlight problems which could be avoided with careful stakeholder mapping and their continuous engagement in the planning, implementation and monitoring processes (Yüksel et al., 1999). Several decades after raising such issues, the inscription of Iznik still reflects a centralized planning system, with apparent consequences for sustainability priorities (Human, 2015; Tosun & Jenkins, 1996; Yüksel et al., 1999). In the case of Iznik, similar to numerous other WHS nominees, residents have not been central in the decision-making processes that informed the city's nomination file. On the other hand, some effort to embrace residents during the inscription process is evident.

Findings suggest that the current focus on Iznik by different stakeholders is heritage conservation. The local administration aspires for tourism activity to become a valuable tool in the economic regeneration of the city, yet there is an apparent lack of coordination among stakeholders that is acknowledged by the administrators; an observation that raises questions about the inscription's economic sustainability. One of the underlying reasons for the present discord stems from the lack of a strategic management plan for Iznik that balances the requirements of urban development, heritage conservation, and local administration, and which embraces different stakeholders. The existence of such a plan does not effectively diminish the conflicts between local and national stakeholders (Lo Piccolo & Todaro, 2014) but it is a potential asset in forming a long-term vision for a place. In this context, different stakeholders set different priorities (Li et al., 2012), as exhibited in the example of the developmental efforts in Iznik versus the conservation mantra of the FMU. The IM representative's aspiration to an alternative course of history, in which the fortress would become uninhabited through developmental restrictions, that would gradually build a new city in the periphery to execute archaeological studies in the fortress, also illustrates the differences of opinion among administrative stakeholders in Iznik regarding its future, both in terms of its residents and its possible inscription as a WHS.

The tangible heritage assets in Iznik are at the forefront of every conversation we had about the city, and the current priority is the conservation of these assets and the city's inscription as a WHS. The comparative evaluation of the PEC administrator of Iznik and European WHSs reflects the perceived uniqueness of the city: "[they] just showed me cities [inscribed] in UNESCO in Europe. Compared to Iznik, they have nothing." Notwithstanding the cultural value embedded in these assets, the derelict condition of some of the tangible assets is also criticized by local community members; attributed to a local elderly community member, "[they should] either build or burn these castles [in Iznik], I don't want to see this filth." Our findings indicate that the administrative informants prefer Iznik to propose an experience instead of a product (Line & Runyan, 2014; Vargo & Lusch, 2004) which would transcend the visitors' activity from a voyeuristic endeavor into an experiential one.

Furthermore, increasing tourism activity as a 'smokeless industry' is viewed as a potential tool to minimize emigration from Iznik. The PEC administrator attributes this to the community's inability to increase economic output in traditional agricultural practice while evaluating the absence of traditional industry as a positive aspect for Iznik and its community. Despite the local administrators' experience with the community's disinterest in support of increasing the tourism activity in Iznik, PEC attaches a paramount importance to training official tourist guides since the absence of information and tools to engage visitors in the city results in many of them not becoming overnight visitors in Iznik. Creating the necessary human capital in Iznik to narratively integrate the city's history with the tangible assets is one way to overcome this limitation. In the words of the IM representative:

> *"[they took us] to the French Parliament. [The guide] talks about the picture on the wall, on and on, for 20 minutes. Then, she explains the stone. ... Sure, this makes no sense to us, this stone, our Iznik is filled with such stones."*

Our discussions on the conservation efforts in the city point towards an uninformed community with regard to the cultural heritage in Iznik. As one of the archaeologists explains, there is a treasure hunting interest in Iznik, partly reinforced through the limited scope and protection of the archaeological studies in the city. The difficulty in conservation efforts is also highlighted by the PEC administrator through a personal anecdote:

> *"[...] the guy says he dug up inside the [Iznik] museum[!] The treasure hunter says 'it may be behind this stone' [and digs] without any considerations to the stone's value."*

Similar conflicts also arise for the intangible heritage in Iznik, which is viewed by our respondents as a primary advantage for Iznik with regard to tourism, and it is also negatively affected by clashes between stakeholders, with the most important example being the çini. Our respondents outline the gradual decline in çini production in Iznik, primarily due to the lack of an official entity to coordinate çini masters. According to the IM representative, most of the çini products sold in Iznik are actually made in Kutahya, a province in central Turkey. This dependence, also a potential outcome of tourism development (Buhalis, 2000), is already occurring, and a potential increase in tourism may further amplify the degeneration of çini as a local art form and an important commodity. In the case of Iznik, this dependence on imports is attributed to the fact that "[you] cannot unify five masters in Iznik. [...] Then five people come from Istanbul, start a foundation, sell çini to the world." Our çini master interviewee also points towards the industry stakeholders' inability to cooperate.

Visitation patterns of the Christian heritage in Iznik as a faith tourism destination is also noted to be declining, yet this decline is attributed to the overall perceptions of visitor safety in Turkey and not to a governance-related paradigm shift, nor to a resistance of the local community. In the cases of both tangible and intangible assets in Iznik, the discoordination between administrative and community stakeholders is apparent, which inevitably influences the trajectory of conservation efforts in Iznik to become a WHS, its tourism prospects, its spatial planning framework, and as an overarching construct, its sustainability objectives.

Iznik, currently, does not clearly portray an inclusion of a sustainability perspective to ongoing efforts in Iznik to become a WHS. While this aim ensures an allocation of increased financial resources by the city (Wuepper & Patry, 2016), there are also questions raised in terms of the economic sustainability of such an allocation. Intertwined with the centralized decision-making system in Turkey, Iznik is an example of heritage by designation (Rautenberg, 1998), with a limited number of institutions and experts working on the heritage aspect of the city (Smith, 2006), and with an even greater number of stakeholders involved in non-heritage-related issues concerning Iznik. Despite the willingness of heritage managers to include local actors in the decision-making process, particularly in relation to the UNESCO procedure, this lengthy and cumbersome process may evolve in its foreseeable course and disregard the residents' and other local stakeholders' opinions at the end (Kavaratzis, 2017). These stakeholders are most affected by strategic planning initiatives due to their geographic proximity, and their expectations from Iznik as a living city.

The juxtaposition of heritage and a living city, in Turkey, implies a multitude of governmental and local institutions responsible for the city's urban fabric (e.g., Somuncu & Yiğit, 2007). From a spatial planning perspective, development within the fortress is limited due to its protected urban site status, whereas Iznik accommodates its increasing population in the northern and southern sections of the fortress. Given the most recent amendments to the urban development plan, regional administrators active in Iznik, i.e., the IM and the BMM, are pushing the limits of the natural borders in Iznik to exploit the land area for tourism and cultural purposes. This is visible in the recent changes made on the 1/1000 urban development plans concerning the shoreline of Iznik Lake in 2017 which allocates the lake's shore (on which development is restricted by national shoreline regulation) to development of cultural and touristic facilities.

The UNESCO procedure's impact on environmental sustainability and the spatial planning of the city remain enigmatic to administrative stakeholders. This process, being at the very early stages of exploration, delegates the decision-making on the spatial characteristics of the city's surrounding area to local administrators, who aim not only to improve locals' quality-of-life but also consider economic improvement in the city. In order to accommodate both priorities, the section of the city outside the fortress that hosts the overwhelming majority of the residents is planned to grow vertically, whereas the scarcely populated areas around Iznik are primarily considered to become visitor oriented. When one combines this spatial planning perspective with the perceived importance of transforming a nearby hill to a panoramic viewing deck and building an eco-tourism facility in the mountainous region surrounding the city to boost and diversify Iznik's tourism prospects, it is not unimaginable to see these areas' eventual inclusion into the urban development plans.

This may trigger an exponentially growing pressure on the city's heritage character since there are cultural heritage assets dispersed intermittently throughout the greater Iznik area. Hence, a coordination between administrators and heritage experts is a valuable instrument for determining the borders of the spatial planning that would both benefit the local residents and foster the integrity of the cultural landscape surrounding Iznik (Papageorgiou, 2015).

The material and immaterial attributes of locations shape residents' sense of place (Campelo et al., 2013) and their actions influence the characteristics of a place (Cresswell, 2014), which imply that locals are integral to all planning initiatives so that the social fabric is preserved. While there is a conservationist stance with regard to the heritage character of the city, the requirements for increased living standards for residents, that boost development mechanisms, are more in line with the necessities imposed by the vision of culture-led tourism development in Turkey (Gunay, 2010; Ozturk & van Niekerk, 2014). In Iznik, as a place in perpetual change (Warnaby & Medway, 2013), the residents are integral parts that maintain a living city and are representatives of the culture embedded in it (Braun et al., 2013; Kavaratzis, 2004). However, as the IM representative notes, "Iznik doesn't have an identity" that would serve the objective to achieve local control by engaging and empowering local communities (UNWTO, 2013). In this perspective, coordination between different stakeholder groups is essential for the sustainability objectives in various planning initiatives, including tourism and urban development.

Implications, limitations and future research

The umbrella provided by the UNESCO procedure is an important asset for the conservation of cultural heritage assets in Iznik and should ideally be interlinked with sustainability. Yet, the apparent disaccord between the experts managing the UNESCO front of Iznik (Rautenberg, 1998; Tweed & Sutherland, 2007), the regional administrators (Lo Piccolo & Todaro, 2014) and residents is one of the main liabilities of Iznik. While tourism is not a direct outcome of inscription (Jimura, 2011; Poria et al., 2013), our findings suggest that experts working on the Iznik project primarily consider the conservation of the assets a top priority, which will ipso facto result in tourism development that would both foster the conservation efforts and generate a substantial financial income to local residents. This would influence how locals perceive tourism activity (Milman & Pizam, 1988) and could contribute to sustainability through social equity, if tourism income is to be utilized to support social programs.

Tourism planning in Iznik takes place without actively engaging another central stakeholder group, the tourists. Drafting of the new legislation in Iznik may prioritize the welfare of the residents but there is a widening gap between the regulations and the heritage character of the city. In this regard, the previous experience that local administrations have with heritage sites and UNESCO in the governance structure of Turkey could be utilized to minimize the negative social impacts of inscription as a WHS (Wang, 2012). The inscription process could be developed upon strategies that embrace these 'largely overlooked' stakeholders, i.e. the residents and the tourists. In fact, taken the recent developments in co-creating the tourism experience, one of the main priorities for future research and action could be how co-creation can be inspired in the case of heritage cities and the case of Iznik in specific.

Turkey is not a stranger to such events with negative social impact and the development agenda in Turkey's historic areas often trigger processes of social transformation in the pursuit of development (Dinçer, 2011; Ergun, 2004; Kocabas & Gibson, 2011). Thus, Iznik's inscription process and the administrative bodies that govern it would benefit from evaluating existing WHSs in Turkey. For example, in Cappadocia, one of the first WHSs in Turkey, whose foreseeably unsustainable tourism development trajectory was suggested as early as the 1990s (Tosun, 1998), subsequent economic dependency on tourism inhibits residents from thoroughly acknowledging the social and environmental costs (Özel & Kozak, 2017) and endangers the area's intangible cultural richness. A closer cooperation of Iznik's FMU with that of Bursa, which has been instrumental in the inscription of Khans in central Bursa and the nearby Cumalikizik village in 2014 and has experience in differing levels of tourism development in different socio-cultural systems within the WHS, would be beneficial due to the sustainability challenges in the tourism development in the latter (Kuzuoglu et al., 2018), but also for addressing the interrelations between tourism development and heritage conservation.

Under advisement of the comparatively large size of Iznik, it is imperative to balance the diversity among different stakeholder groups. To this end, in the drafting of the management plan for Iznik, the necessary spatial interventions (both in relation to the heritage aspect of the city, and in accordance with the residents' expectations from their city) may benefit from building upon inclusive stakeholder mapping and then be discussed from early stages on, so that conflicts that may arise throughout the process can be minimized (Lo Piccolo & Todaro, 2014). Tourism's impact on cities is well-known, yet management plans for WHSs often fail to investigate the potential outcomes of tourism development, primarily due to the fact that these management plans are preoccupied with heritage conservation, meaning that sustainability and the social and cultural impacts are not prioritized.

As explained by Human (2015:174) from an ethnographic fieldwork in Catalhoyuk in central Turkey "Ministry officials' approach to stakeholder consultation, [...], aims to reinstate a subordinate, disempowered role for local actors in the site management planning process." Despite the perspective of the central governance mechanisms, Human's (2015) account on the disconnect between the local and the national, may be addressed in Iznik by involving as many stakeholders as possible prior to the preparation of the management plan. This may not only be beneficial in the planning efforts of the cultural and spatial landscape in Iznik, but it would also be helpful in determining the expectations of tourists and local residents in terms of conservation efforts and a sustainable tourism development (UNWTO, 2013). In this perspective, the management plan is the blueprint for the future development of the city if it becomes a WHS and virtually every planning initiative in and around Iznik becomes dependent on it.

Heritage sites are inherently linked to tourism development as one of Turkey's national objectives. While the central governance structure in Turkey currently has minimal financial allocation for heritage sites, the recent decentralization effort of state-sponsored (or subsidized) investments through RDAs (Alvarez et al., 2014) is a valuable instrument for incentivizing increased investments in heritage cities like Iznik. Despite the efforts of minimizing the bureaucratic burden on local authorities, the existing hierarchical ladder in different contexts of governance throughout Turkey inhibits the local administrations to work with a certain level of autonomy. Thus, the heritage character of Iznik and the management plan that oversees it needs to be central in determining the investment framework since the absence of guiding principles in the development planning phase may lead to undesired social problems and an imbalance in economic structures.

Heritage is only a single part of Iznik, and, arguably, this part has received the overwhelming majority of scholarly interest in the city. Yet, our analysis reiterates the importance of consensus-building in the complex structure of a living city with ancient heritage. Iznik possesses a variety of resources that may be leveraged to differentiate the city from its competitors (Ross et al., 2017). In order to position these resources to regenerate the local economy while serving sustainability objectives, adequate monitoring and auditing mechanisms are required starting from the early stages of strategic planning (Damurski & Oleksy, 2018). In this vein, the discoordination between different administrative stakeholders needs to be unified under a common vision and strategic plan.

The current UNESCO procedure, not disclosed to the general public to its full extent, inhibited us from concentrating on the perspectives of the local residents. Sustainability of their cities as tourism destinations affect residents to varying extents, and as exemplified by heritage cities experiencing over-tourism, residents' quality-of-life may be endangered by continuous, imbalanced increase of tourism activity (Seraphin et al., 2018). In order to foresee potential societal barriers that may arise as the city and its tourism activity develop, the resident characteristics and socio-cultural attributes in Iznik need to be studied to a deeper extent. In this context, an in-depth analysis of the lost and/or prevailing cultural assets as perceived by the residents, continuing the work of Yücel and Salt (2018), would complement the findings of this research.

Despite a possible improvement in terms of coordination between different administrative branches in Iznik over the course of the UNESCO procedure, this process' inevitable influence on the cultural and spatial landscape of Iznik requires continuous monitoring both by regional or national authorities and by non-governmental pressure groups (Li et al., 2012) for the potential ramifications on the resident community. Last, but not least, the present study acknowledges the importance of tourism development to local administrators, but recommends that the administrative organs responsible for the planning and management of tourism should holistically approach the notion of tourism development in Iznik by co-operating with various administrative entities to work towards improving community well-being, safeguarding the cultural heritage assets and environmental conditions in and around Iznik while increasing the economic benefits accrued through tourism activity.

References

Akbaygil, I., Inalcık, H. & Aslanapa, O. (eds.) (2003) *Iznik Throughout History*, Istanbul: Türkiye İş Bankası Kültür Yayınları.

Alvarez, M. D., Ertuna, B., Ünalan, D. & Hatipoğlu, B. (2014) 'The role of regional agencies', *Tourism as an Instrument for Development: A Theoretical and Practical Study*, London: Emerald Group Publishing, pp. 127-147.

Andereck, K. L., Valentine, K. M., Knopf, R. C. & Vogt, C. A. (2005) 'Residents' perceptions of community tourism impacts', *Annals of Tourism Research*, 32 (4), 1056-1076.

Ashworth, G. J. & Page, S. J. (2011) 'Urban tourism research: Recent progress and current paradoxes', *Tourism Management*, 32 (1), 1-15.

Aslanapa, O. (2003) 'The origin of the Ottoman keramic art and its development', in Akbaygil, I., Inalcık, H. & Aslanapa, O. (eds.), *Iznik Throughout History*, Istanbul: Türkiye İş Bankası Kültür Yayınları.

Atakuman, Ç. (2010) 'Value of heritage in Turkey: History and politics of Turkey's World Heritage nominations', *Journal of Mediterranean Archaeology*, 23 (1), 107-131.

Bayraktar, M. (1994) 'Davud-i Kayseri', *İslam Ansiklopedisi*, Ankara: Türkiye Diyanet Vakfı Yayınevi, pp. 32-35.

BMM. (2015) Iznik/Nicaea on Its Way to Become a UNESCO World Heritage Site, 2-4 October, Iznik, Bursa, Turkey.

Braun, E., Kalandides, A., Kavaratzis, M. & Zenker, S. (2013) 'My city – my brand: the different roles of residents in place branding', *Journal of Place Management and Development*, 6 (1), 18-28.

Buhalis, D. (2000) 'Marketing the competitive destination of the future', *Tourism Management*, 21 97-116.

Butler, R. W. (1980) 'The concept of a tourist area cycle of evolution: implications for management of resources', *The Canadian Geographer/Le Géographe canadien*, 24 (1), 5-12.

Campelo, A., Aitken, R., Thyne, M. & Gnoth, J. (2013) 'Sense of place: The importance for destination branding', *Journal of Travel Research*, 53 (2), 154-166.

Cresswell, T. (2014) *Place: An introduction*, West Sussex, UK: Wiley-Blackwell.

Damurski, Ł. & Oleksy, M. (2018) 'Communicative and participatory paradigm in the European territorial policies. A discourse analysis', *European Planning Studies*, 26 (7), 1471-1492.

Dinçer, İ. (2011) 'The impact of neoliberal policies on historic urban space: Areas of urban renewal in Istanbul', *International Planning Studies*, 16 (1), 43-60.

Dupagne, A., Ruelle, C., Teller, J. & Cornélis, B. (2004) *SUIT: Sustainable Development of Urban Historical Areas Through an Active Integration Within Towns*, Luxembourg: Office for Official Publications of the European Communities.

Ergun, N. (2004) 'Gentrification in Istanbul', *Cities*, 21 (5), 391-405.

Falser, M. (2015) *Cultural Heritage as Civilizing Mission*, Switzerland: Springer International.

Foss, C. (2003) 'The walls of Iznik 260-1330', in Akbaygil, I., Inalcık, H. & Aslanapa, O. (eds.), *Iznik Throughout History*, Istanbul: Türkiye İş Bankası Kültür Yayınları, pp. 249-262.

Glaser, B. G. & Strauss, A. (1967) *The Discovery of the Grounded Theory: Strategies for Qualitative Research*, New York: Aldine.

Gunay, Z. (2010) 'Conservation versus Regeneration?: Case of European Capital of Culture 2010 Istanbul', *European Planning Studies*, 18 (8), 1173-1186.

Harrill, R. & Potts, T. D. (2003) 'Tourism planning in historic districts: Attitudes toward tourism development in Charleston', *Journal of the American Planning Association*, 69 (3), 233-244.

Human, H. (2015) 'Democratising world heritage: The policies and practices of community involvement in Turkey', *Journal of Social Archaeology*, 15 (2), 160-183.

IM. (2005) Uluslararası İznik Sempozyumu (International Iznik Symposium), 5-7 September, İznik, Bursa, Turkey.

Jimura, T. (2011) 'The impact of world heritage site designation on local communities – A case study of Ogimachi, Shirakawa-mura, Japan', *Tourism Management*, 32 (2), 288-296.

Joy, C. (2016) ''Enchanting town of mud': Djenné, a World Heritage Site in Mali', in De Jong, F. & Rowlands, M. (eds.), *Reclaiming Heritage: Alternative Imaginaries of Memory in West Africa*, London: Routledge, pp. 145-159.

Kahraman, S. A. & Dağlı, Y. (eds.) (2013) *Günümüz Türkçesiyle Evliya Çelebi Seyahatnamesi (Evliya Çelebi's Travelogue in Today's Turkish)*, İstanbul: Yapı Kredi Yayınları.

Kavaratzis, M. (2004) 'From city marketing to city branding: Towards a theoretical framework for developing city brands', *Place Branding*, 1 (1), 58-73.

Kavaratzis, M. (2017) 'The participatory place branding process for tourism: Linking visitors and residents through the city brand', in Bellini, N. & Pasquinelli, C. (eds.), *Tourism in the City*, Switzerland: Springer, pp. 93-107.

Kocabas, A. & Gibson, M. S. (2011) 'Planned gentrification in Istanbul: The Sulukule renewal area 2005-2010', *International Journal of Sustainable Development and Planning*, 6 (4), 420-446.

Kuzuoglu, S., Karagöz, A. Ö., Sürmeli, P. & Hatipoğlu, B. (2018) 'Historical roots of tourism development in Cumalikizik: Balanced or ill-fated?', International Conference on Tourism, 27-30 June, Kavala, Greece.

Landorf, C. (2009) 'Managing for sustainable tourism: A review of six cultural World Heritage Sites', *Journal of Sustainable Tourism*, 17 (1), 53-70.

Li, T. H. Y., Ng, T. S. & Skitmore, M. (2012) 'Conflict or consensus: An investigation of stakeholder concerns during the participation process of major infrastructure and construction projects in Hong Kong', *Habitat International*, 36 (2), 333-342.

Line, N. D. & Runyan, R. C. (2014) 'Destination marketing and the service-dominant logic: A resource-based operationalization of strategic marketing assets', *Tourism Management*, 43 91-102.

Lo Piccolo, F. & Todaro, V. (2014) 'From planning to management of cultural heritage sites: Controversies and conflict between UNESCO WHL management plans and local spatial planning in south-eastern Sicily', *European Spatial Research and Policy*, 21 (2), 47-65.

Lowry, H. W. (2003) 'Ottoman Iznik (Nicaea): Through the eyes of traveler & as recorded in administrative documents, 1331-1923', in Akbaygil, I., Inalcık, H. & Aslanapa, O. (eds.), *Iznik Throughout History*, Istanbul: Türkiye İş Bankası Kültür Yayınları, pp. 135-174.

Milman, A. & Pizam, A. (1988) 'Social impacts of tourism on Central Florida', *Annals of Tourism Research*, 15 191-204.

Özel, Ç. H. & Kozak, N. (2017) 'An exploratory study of resident perceptions toward the tourism industry in Cappadocia: a Social Exchange Theory approach', *Asia Pacific Journal of Tourism Research*, 22 (3), 284-300.

Ozturk, A. B. & van Niekerk, M. (2014) 'Volume or value: A policy decision for Turkey's tourism industry', *Journal of Destination Marketing & Management*, 3 (4), 193-197.

Papageorgiou, M. (2015) 'Planning practices for the protection of cultural heritage: Lessons learnt from the Greek UNESCO sites', *European Spatial Research and Policy*, 22 (2), 111-125.

Pendlebury, J., Short, M. & While, A. (2009) 'Urban World Heritage Sites and the problem of authenticity', *Cities*, 26 (6), 349-358.

Poria, Y., Reichel, A. & Cohen, R. (2013) 'Tourists perceptions of World Heritage Site and its designation', *Tourism Management*, 35, 272-274.

Rautenberg, M. (1998) 'L'émergence patrimoniale de l'ethnologie: entre mémoire et politiques publiques', in Poulot, D. (ed.) *Patrimoine et Modernité*, Paris: L'Harmattan, pp. 279-289.

Ross, D., Saxena, G., Correia, F. & Deutz, P. (2017) 'Archaeological tourism: A creative approach', *Annals of Tourism Research*, 67, 37-47.

Ryan, G. W. & Bernard, H. R. (2003) 'Techniques to identify themes', *Field Methods*, 15 85-109.

Seraphin, H., Sheeran, P. & Pilato, M. (2018) 'Over-tourism and the fall of Venice as a destination', *Journal of Destination Marketing & Management*, 9, 374-376.

Smith, L. (2006) *Uses of Heritage*, New York: Routledge.

Somuncu, M. & Yiğit, T. (2007) '*Türkiye'deki Dünya Mirası Alanları'nın Güncel Durumlarının Saptanması Raporu: Görme Milli Parkı ve Kapadokya Kayalık Sitleri (Report on Determining the Current Status of World Heritage Sites in Turkey: Göreme National Park and the Rock Sites of Cappadocia)*', report prepared by UNESCO Turkey National Commission.

Stefanidou, V. (2003) Nicaea (Antiquity), Available: http://asiaminor.ehw.gr/forms/fLemmaBodyExtended.aspx?lemmaid=8503&boithimata_State=&kefalaia_State=#chapter_3 [Accessed 26.01.2020].

Tosun, C. (1998) 'Roots of unsustainable tourism development at the local level: The case of Urgup in Turkey', *Tourism Management*, 19 (6), 595-610.

Tosun, C. & Jenkins, C. L. (1996) 'Regional planning approaches to tourism development: the case of Turkey', *Tourism Management*, 17 (7), 519-531.

Tweed, C. & Sutherland, M. (2007) 'Built cultural heritage and sustainable urban development', *Landscape and Urban Planning*, 83 (1), 62-69.

UNESCO. (1972) *Convention Concerning the Protection of the World Cultural and Natural Heritage,* report prepared by United Nations.

UNESCO. (2014) İznik, Available: https://whc.unesco.org/en/tentativelists/5900/ [Accessed 13.07.2018].

UNWTO. (2013) Sustainable Tourism for Development Guidebook, Available: http://whc.unesco.org/sustainabletourismtoolkit/sites/default/files/13.%20UNWTO%20%282013%29%20Sustainable%20Tourism%20for%20Development%20Guidebook.pdf [Accessed 29.01.2020].

Vareiro, L. M. d. C., Remoaldo, P. C. & Cadima Ribeiro, J. A. (2013) 'Residents' perceptions of tourism impacts in Guimarães (Portugal): A cluster analysis', *Current Issues in Tourism,* 16 (6), 535-551.

Vargo, S. L. & Lusch, R. F. (2004) 'Evolving to a new dominant logic for marketing', *Journal of Marketing,* 68 (1), 1-17.

Vecco, M. (2010) 'A definition of cultural heritage: From the tangible to the intangible', *Journal of Cultural Heritage,* 11 (3), 321-324.

Wang, S.-Y. (2012) 'From a living city to a World Heritage City: Authorised heritage conservation and development and its impact on the local community', *International Development Planning Review,* 34 (1), 1-17.

Warnaby, G. & Medway, D. (2013) 'What about the 'place' in place marketing?', *Marketing Theory,* 13 (3), 345-363.

Wuepper, D. & Patry, M. (2016) 'The World Heritage list: Which sites promote the brand? A big data spatial econometrics approach', *Journal of Cultural Economics,* 41 (1), 1-21.

Yücel, S. D. K. & Salt, E. (2018) 'Detection of cultural landscapes changes in the historical process: The case of Bursa İznik', *Planlama,* 38 (1), 40-55.

Yüksel, F., Bramwell, B. & Yüksel, A. (1999) 'Stakeholder interviews and tourism planning at Pamukkale, Turkey', *Tourism Management,* 20 351-360.

Yüksel, F., Bramwell, B. & Yüksel, A. (2005) 'Centralized and decentralized tourism governance in Turkey', *Annals of Tourism Research,* 32 (4), 859-886.

Part 4. Technology, residents and overtourism

10 How digital strategy increases overtourism – the case of Barcelona

Stephane Bourliataux Lajoinie, Josep Lluis del Olmo Arriaga and Frederic Dosquet

The objectives of this chapter are to:

- ☐ Analyse how digital communication can escape the control of DMOs.
- ☐ Check how two forms of digital communication (promotion and social networks) generate a self-phenomenon of spreading through internet.
- ☐ Check how recommendation platforms promote overtourism in very small local areas.
- ☐ Have an open discussion about communication co-creation at the edge of DMO and tourists' point of view.

Keywords: Digital promotion, social networks, overtourism, review platforms.

Introduction

The term 'overtourism' made the headlines in 2017 to denote the antithesis of a fair, well-planned form of tourism. An overtourism destination is a destination perceived as the victim of uncontrolled flows of tourists. Some authors have contributed to analysing the first stages of this perception. Boissevain (1996) published 'Coping with tourists: European reactions to mass tourism', in which he analysed how a massive flow of tourists in Malta generated negative side effects and a total dependence on touristic economy. Tyler et al. (1998) offered one of the first analyses of ways to control tourism flow from a sustainable point of view. Bosselman et al. (1999) published a pioneering analysis on the relationships between hosts and tourists. The paper discusses the risks and benefits of growth in tourism and the need for long-term management to avoid overexploitation of tourist destinations. These research projects laid the foundations for reflection on sustainable tourism and were the starting point for studying the risks of overtourism. Fyall and Garrod (1998) discussed the impacts of overtourism on heritage sites. On the one hand, the phenomenon contributes to a short-term economic boost; on the other hand, it impacts negatively on the way of life of local communities. Research on the topic of are centred on three main areas:

- Excessive burden on transit stations (airports, ports, stations), the problem of managing the flow of movement (bottlenecks, parking), and the management of other external factors related to the influx of people (drinking water supply, and waste management). This component also includes management of infrastructure costs based on local taxes that are, by definition, paid by tourists only indirectly (tourist taxes on hotel stays) and in their vast majority paid by residents.
- Covering interactions between tourists and locals. Great tourist pressure requires efforts in cultural acceptance by both sides, as well as showing respect for local customs. Several studies report a very negative perception of the noisy party behaviour shown by some tourists (Martin et al., 2018). The flow of tourists also increases the volume of population movement in daily life, making transportation and waiting times a problem in some places.
- Economic impact linked to tourism attractiveness. While the tourism economy supports many tourist structures, it also has a direct impact on the cost of housing and the economic value of certain neighbourhoods. First studied in 2011, residents have identified tourism as the

most serious problem in Barcelona (Milano, 2017). In the space of six years, the negative perception of tourists by Barcelonans multiplied by a factor of 23 (source: Barcelona City Council, 2018). Residents now feel that the problem of overtourism is more important than working conditions, economic problems and the state of municipal infrastructure and services.

The reasons for this overtourism are numerous and their impact is usually crossed. However, we find one constant at the root of the feeling of overtourism: a rapid, uncontrolled increase in the flow of tourists to a destination. Recently many authors have studied the perception of overtourism among the local population to try to determinate the acceptable level of tourism pressure (Koens et al 2018; Gonzalez et al. 2018; Kuščer & Mihalič 2019) but only recently have some of them studied the impact of the internet on these processes (Alonso-Almeida et al. 2019, Gretzel 2019, María-del-Mar 2019).

Spanish north-east coast: from an economically stricken region to the temple of mass tourism

The cities' own approach and development of a tourism strategy may be problematic too. The tourism economy is a strong lever for cities whose industrial economic environment is in decline. Despite the hard conditions for workers, tourism can bring new jobs that are strongly non-relocatable (Walmsley, 2017). By way of example, the reconstruction in the 1980s of the northern port districts of Barcelona for the creation of the Olympic facilities was an opportunity to convert old abandoned factories into resources for the tourism economy. For their part, many cities have conducted in-depth studies confirming the positive impact of pursuing mass tourism (Kock et al. 2018). The creation of structures such as destination marketing organisations (DMO) reflects the public and private will to monetise tourism (Koens et al. 2018). DMOs are public structures designed to promote the tourist attractiveness of a city or geographical area, and their main mission is to highlight the characteristics of a place and attract tourists (Mak, 2017). At the same time, taking advantage of the economic manna of tourism, cities like Barcelona have undertaken major construction work to rehabilitate neighbourhoods and develop cultural spaces. The number of tourists visiting Barcelona has doubled since the year 2000 to nearly 8.8 million people making just over 18 million overnight stays. This figure does not include the 21 million 'temporary' tourists, for example those visiting on cruises, who make only one-day stop-offs in the city.

However, the impact of digital communication is especially noticeable in the case of accommodation rented between individuals. Barcelona City has 13,555 private rental listings and Airbnb has 17,369 listings for the city and its surrounding area. This compares to the 72,000 hotel nights recorded in 2017.

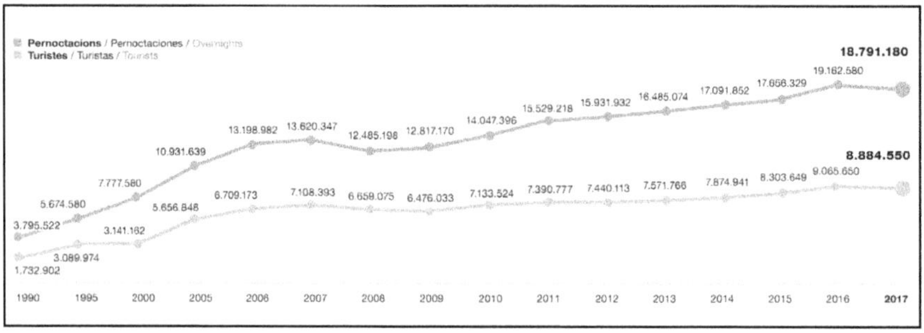

Figure 10.1: Number of tourists and nights in Barcelona. Source : Barcelona City Council (2018).

The opening up of the city to this mass tourism has, in turn, led to an evolution in both physical and digital means of communication, which has helped Barcelona to be named as "European Capital of Innovation", or iCapital, in 2014, generating even more tourist appeal.

From digital communication opportunities to loss of control

New technologies have had a significant impact on the evolution of choice of tourist destination (Steen Jacobsen & Mumar 2012); sharing platforms and social networks have completely changed the relationship between the image of a site driven by DMOs and the image managed and offered to tourists via other tourists. Key elements in user behaviour now range from the search for a holiday resort and organising their stay (Buhalis & Law, 2008) to advice and feedback from digital tools (the internet, mobile applications). The main users of this type of application are tourists who are planning to visit the city in a few hours and who, having little time, need a product offering a guided route that is easy to consult and browse. This type of informal tourism, hurried and concentrated in the top tourist attractions, reinforces the perception of overtourism among locals.

The use of digital media creates an emotional connection between consumers and tourism entities. Due to an element of mimicry, this contributes to an increase in tourist pressure in areas that are already well known. The reply to a simple Google search for the keywords "a big YES to..."

or "dream stay in..." usually sees the same cities appear automatically: Venice, etc. While it is true digital communication is a great opportunity for DMOs, it is also true we must also take into account its dark side. Communication is hard to track and escapes the authorities. Tourist reviews and attractive blogs and forums create a line of communication that is out of the city's control.

What began as opportunity, sharing experiences on the web is now a problem in the fight against overtourism. "The interest of the web no longer lies in the content it offers but in the links of the social connections it generates, connections between people that create a social fabric" (Hossler et al. 2014).

This has several consequences; the first is the change in the classic hierarchy of effects in the paid media, owned media and earned media approach (Harrison 2013; Lovett & Staelin 2016). Till the loss of control by DMOs, the promotion strategy of big tourist cities was managed through these four main way of communication. The central problem was to keep hold of the official strategy. But in our opinion things have changed now and DMOs need to manage another mix of communication, where earned promotion exceeds controlled promotion.

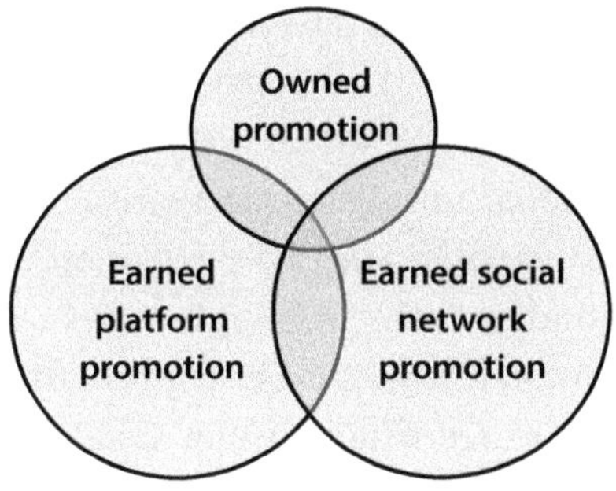

Figure 10.2: The new deal of promotion management for DMOs. Source: Authors' elaboration.

Owned promotion carried out by DMOs is now decreasing rapidly; the destination itself generates so many internet interactions than we will focus on these three dimensions in the mix of communication.

Digital promotion

Figure 10.3 shows the Google Trends for "visit Barcelona" between 2005 and 2020, i.e. the numbers of searches made for the term over that period. Since 2005, the requested keywords have doubled in Google searches. Despite some seasonal variations easily seen in Figure 10.3, the keywords increase systematically year on year. This first image shows how Visit Barcelona has become a more and more popular search term.

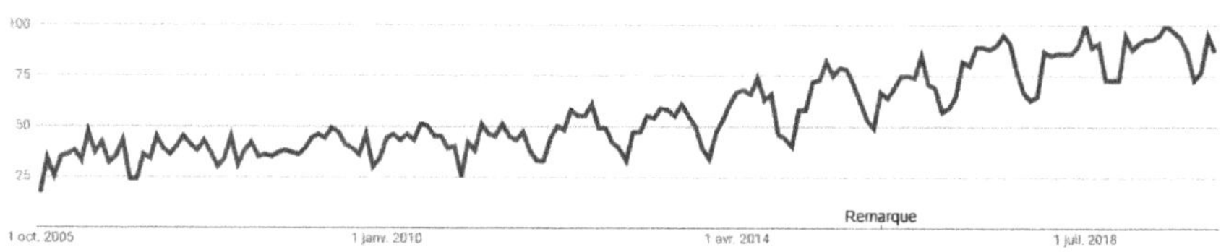

Figure 10.3: Evolution of "Visit Barcelona" during the last 15 years. Source: Google Trends.

Overtourism became a real issue as early as 2004, with Barcelona leading the way among destinations complaining about the negative impacts of the phenomenon. E-commerce has enabled the services economy to flourish; a service that is inherently intangible, perishable, and heterogeneous (Lovelock & Wirtz, 2011), it adapts perfectly to the constraints of instanta-neity of the web. Tourism companies and DMOs have been able to develop a profit management strategy by offering prices adapted on a daily basis according to the filling rates of booking platforms. Although this practice was for a long time reserved for airlines, it is now the basic logic behind booking platforms such as Booking.com. The use of mobile applications such as Depart Demain ('leave tomorrow') has further reinforced this phenomenon by offering users particularly low prices if they agree to buy unsold tickets within 72 hours of departure. This true tourism opportunity has been implemented; the price variable and the perception of doing a good job replaces planning (Schindler, 1989). A simple "visit Barcelona" in Google returns more than 280 million results (4 September 2019). Organising a trip in Barcelona is easier than anywhere else in Catalonia, but going to the simplest deal puts all tourists in the same place. Overtourism was started by tour operator policies to promote the most profitable and easiest ways to book trips on the internet. The official site of www.barcelonanavigator.com recommends 31 applications to help tourists during their visit. These applications were downloaded more than 600,000 times during 2017, rising to 800,000 in 2019 and will probably surpass 1 million in the near future, when mass tourism recovers after the Covid-19 downturn.

The use of ICT also permits privileged access to the data of Internet users. Although the notion of privacy and research in this field has been growing for some years, consumers are still largely unaware of the data they leave on the web or via their smartphones. This data forms the basis of a data economy, and tourism operators are well-versed in ways to make use of such data. It is thus possible to offer ultra-specialised offers to segments of the population as defined by their expectations in terms of price, location, flight time, additional activities and so on. By capturing browsing data, operators can adapt their offerings in real time to highlight the criterion that will be most relevant in triggering the act of purchase. Since this data is based on behavioural analysis of users of the website or the mobile application, their probability of effectiveness is particularly high. By presenting the same offers but optimally packaged and presented, thanks to the data collected, tour operators optimise the rate of conversion of prospects into customers, increasing sales, particularly for the most popular destinations. This kind of analysis generates overexploited tourist routes in cities. Applications like Monument Tracker recommend the most-used routes, increasing the tourist density in some areas at the expense of others.

If we summarise the digital impact of the only official baseline keywords "visit Barcelona" in 2019 we can find:

- The Google search engine gives 542 million results with the official website barcelonatourisme.com in the two first rankings and Tripadvisor on line three.
- The Twitter social network tag #visitbarcelona was used 412,000 times (or 1,128 times a day)
- The Instagram tag @visitbarcelona is followed by 316,500 people with 1,581 yearly posts.
- The Pinterest page "visitBacelona" is followed by 4,100 people and seen by 6,000 users every month
- In Facebook, more than 118,000 people like the official page.

Social network promotion

Social networks offer an emotional relationship between the brand and consumers of tourist sites. In addition to highlighting commercial offerings and the data captured on the profiles of users, they allow a true structuring of the experience according to the charm of resorts (Hetzel, 2002). The use of Facebook pages or Instagram and Pinterest accounts can highlight a place, or a way of life as imagined by the consumer ("Barcelona by night", etc.). The consumer is thus attracted through two dimensions of the postmodern approach, as defined by Decrop (2008): hyperreality and fragmentation. The hyperreality of consumption is manifested by the consumer's perception that they know the resort before even going there. Videos and images made available on the official social networks of operators are the most attractive assets of tourist sites, as they also ignore any neighbourhood nuisance related to tourist pressure. The reality of the stay is then presented, leaving out any possible nuisances, offering an idealised image of the place that the tourist is looking for. The fragmentation of consumption is reflected by the opportunistic behaviour of tourists, in which the consumer picks the destinations to create their personal image of being a traveller. The image generated by this stay is called on to consolidate one of the facets of the personality of the tourist who makes it. On social networks, social compliance creates identical tourist expectations for millions of tourists (Ayeh et al., 2012). The origin of overtourism can be found in the will to conform to a standardised view of everything, with people taking the same photo from the same angle, etc.

Due to YouTube channels, Instagram and many other bloggers, some destinations and attractions have moved from the shadows to take the limelight in the tourism industry. The contribution of digital media to this evolution of tourism can be analysed in four points: adaptation between the type of service sold and the internet medium; simplified access to user profile data and tracking; affective relationship maintained by brands; and, finally, value co-creation between users. All of these contribute to creating a perception of overtourism among local hosts.

The main contribution of social networks in terms of promoting a site is in their ability to give the illusion of close, personal recommendation (Tuten & Salomon, 2017). The ease of use of social networks allows any tourist to contribute to the promotion of a site by submitting their photos. We will retain here two dimensions proposed by Grerzel (2019) for the contribution of social networks to tourism: a dimension of social conformity, and a dimension of highlighting one's personal life (which Grerzel defined as a panoptic force of social media).

The social compliance dimension takes us directly to the extrinsic and social orientation of Holbrook (1999). Social media has changed the way tourists portray their visits to sites. The search for the selfie in an emblematic place (arms spread at the feet of the Corcovado in Rio, the play of perspective and the finger stretched out on the top of the Louvre pyramid, etc.) are all clichés repeated endlessly by tourists who thus wish to fit into the mould of social conventions of the ideal photo at the site visited (Chen et al. 2010; Yoo et al. 2014). The style of the photograph (reference to a pose taken by an influencer or network celebrity) is just as important as the reference to the place visited (Dinhopl & Gretzel, 2016). Here, the tourist seeks to master the codes of a symbolic language highly codified by influencers (Acott, 2018). As a consequence of this search for social conformity, already highly visible venues are overloaded with tourists in search of a photo that will allow them to imitate influencers and thus increase their own social influence (Chen et al., 2009). Large queues form in front of the square metre dedicated to the ideal selfie, which in turn must look as if it has been improvised by its author. This brings us to the second dimension analysed here: highlighting one's personal life. The use of social networks makes it possible to propagate an idyllic vision of one's life and of one's visit to a tourist venue (Larsen et al., 2007; Gosling et al., 2011; Moon et al., 2016). What tourist would want to put on their social network a photo of Thai beach in Maya Bay with the beach spoiled by litter and dozens of boats waiting to disembark? This enhancement of perfect moments in social exchanges has a powerful effect on tourism marketing: overvaluation, a priori of the tourist experience. The intangibility of the tourist service frequently leads to an overvaluation of the experience to be had (Chon, 1991; Foley & Fahy, 2004) if the user finds on social networks only perfect images posted by previous tourists there, and this sets high expectations for their future trip to the venue. This can be seen both in the desire to visit exactly the same site as that seen in the photo (hence the overtourism process of the most photographed sites) and in the probable disappointment with the number of tourists at the venue, not to mention the constraints external to the organisation, such as the weather.

Online review platforms

Over the past few years, the use of online social networks has become particularly commonplace in the tourism sector. These days, TripAdvisor is used as a major tool in helping select a tourist destination or activity. As highlighted by Paquerot et al. (2011), tourism is an intangible, experi-ential proposition whose physical attributes (atmosphere, hospitality, etc.) are difficult to rate. Thus, reliance on the opinion of other consumers has become more and more common when preparing a trip or visit to a tourist attraction. In fact, the number of people checking online review platforms when selecting a tourist destination is growing (Buhalis & Law, 2008). According to Filieri et al. (2015), 200 million travellers will consult platforms like TripAdvisor when planning a visit to a specific destination (travel, restaurant, hotel, etc.). Thus, better understanding the way these online ratings are perceived, and the way they impact behaviour, is an important issue for many tourist and cultural organisations, whose e-reputation may be affected by this type of platform (Hennig-Thurau et al., 2004 in Filieri et al., 2015).

Equally, this issue is even more important for tourist and cultural attractions located in areas that have a large number of historical attractions. This is true for the city of Barcelona with its highly touristic sites. In such cases, a tourist will use a review platform before (to prepare the trip) and after their visit (Xiang and Gretzel, 2010). The main perceived characteristics of online ratings are:

- The content's perceived trustworthiness (Mauri & Minazzi, 2013; Xu, 2014; Casaló et al. 2015)
- The content's perceived usefulness (Casaló et al. 2011, 2015)
- A diminution in the perceived risk in making a decision (Gretzel, Yoo & Purifoy, 2007)

Perceived trustworthiness is a concept that can be understood in different ways and that depends more on individual judgement than on the inherent characteristics of a source (e.g. trust, authenticity, transparency, competence, integrity; Gurviez & Korchia, 2002; Johnson & Kaye, 2009). It relies essentially on written comments, with importance placed on on reviewer expertise (e.g. Racherla & Friske, 2012) and readability (Korfiatis, Garcia-Bariocanal & Sanchez-Alonzo, 2012). It therefore does not appear to be related to metric evaluations, even though a high rating may mean that the Internet user will lend more weight to the level of trust given to ratings.

Perceived utility is also a subjective assessment as it relates to attitude (positive or negative valence, and greater or lesser intensity) based on how the cost-benefit ratio of a product or service's use is perceived (e.g. Davis, 1989). Thus, the issues raised in written comments may provide information sought by travellers, whereas metric evaluations simply serve to confirm or overturn the overall expected quality of a service provided (reception, hotel, etc.).

Finally, the decision to book an unknown destination or decide to visit a cultural site involves the idea of reducing the risk perception associated with this decision (Gretzel, Yoo & Purifoy, 2007). Individuals naturally seek to identify the positive and negative aspects of a destination they plan to visit (Chen & Uysal, 2002) ; "Online consumer reviews can be considered a form of e-WOM" (Hennig-Thurau et al., 2004 in Filieri et al., 2015). Moreover, the online reviews viewed on official websites, (e.g. TripAdvisor) are perceived as being:

- More trustworthy than reviews posted directly on organisation sites (e.g. Bansal & Voyer, 2000 ; Casaló et al. 2015) ;
- More trustworthy than travel agents and mass media (e.g. Dickinger, 2011)

These official platforms also offer the advantage of standardising the featured measures and their presentation: it is therefore easier for an individual to compare two or more destinations. Of course, the trust placed in reviews applies essentially to written comments. Note that on this subject, Plotkina, Munzel and Pallud (2017) have developed an algorithm that enables an 81% detection rate for false advice. Thus, application of the algorthm makes it possible to limit fraudulent practices in which, for example, hotel managers encourage employees to post negative reviews on competitor sites (Filieri et al., 2015).

As such, other studies have focused on examining the profile of the reviewer (e.g. Lee et al., 2011) and show that those who provide the most opinions on review platforms are those who travel most, and who most negatively evaluate destinations, regardless of gender or age.

Finally, the co-creation of value among consumers is an important source of tourist pressure. This co-creation consists, in our case, of the use of platforms to assess tourist destinations and services. The best known of these platforms is, without doubt, TripAdvisor which, as its name suggests, aims to offer tourists advice. Social recommendations are now an effective lever of persuasion for consumers in the information gathering stages. Studies generally suggest an uncontrollable value co-creation on evaluation platforms (Nicholls, 2010), or a non-discriminating element in view of the natural attractiveness of the tourist sites evaluated (Bourliataux-Lajoinie & Maubisson, 2017). This consumer interest in tourism assessment platforms echoes two other dimensions put forward by Decrop (2008): alone and together, and real and virtual. Alone and together reflects the behaviour of participants in these evaluation networks who will display their independence while having the largest audience possible. Their individual personality needs the evaluation of the group to assert itself. Finally, the real-virtual fusion again blurs the border between the two worlds: the tourist destination is imagined and idealised before the visit, but it will be staged in times shared on the network. Idealised images of the destination reinforce the chimerical construction of the place and encourage other participants (fans, followers, etc.) to reinforce the myth of the idyllic destination. In addition to individual added value (linked to the search for a destination by an individual), the creation of value by consumers also impacts on the social value of the service, in this case, a destination (Belk, 2010; Sigala, 2017), by promoting the attractiveness of the site and the quality of ratings.

Conclusion

The contribution of digital to tourism communication is undeniable, but it would be useful nowadays to explore the behaviour of tourists online and the sharing of information as tools for implementing tourist exit policies. A measure of the dissemination of the digital image of a city or an evaluation of the rating tools like TripAdvisor, would be one way to control the tourist-phobic behaviour while better understanding the expectations of visitors. The co-creation of value by tourists (TripAdvisor ratings or blogs and personal accounts on social networks) is an element that destinations should take into account as part of a tourist exit strategy, in which a city like Barcelona seeks to change its image and especially to regulate its attractiveness. In this particular case, the concept of smart tourism represents a real opportunity for a DMO. Using an app that tries to redirect tourists to a new part of the city may be a wise choice to decrease tourist pressure in popular places. The content of user-generated messages is more effective than traditional advertisements in changing the image of a service (Del Chiappa et al., 2015; Narangajavana et al., 2017). An analysis of the background to this behaviour, the mechanisms of its structuring and its acceptance, in line with the work of Herrero et al., 2017, would help to control tourist-generated content (TGC) better and guide the latter towards repositioning Barcelona's target group. Our research agenda leads us in particular to address these two themes through a study conducted on TripAdvisor's TGC concerning Barcelona, and on the use of mobile services as offering added value to the tourist opportunities perceived by users.

References

Alonso-Almeida, M. D. M., Borrajo-Millán, F. & Yi, L. (2019). 'Are social media data pushing overtourism? The case of Barcelona and Chinese tourists.' *Sustainability*, 11(12), 3356.

Acott, K. (2018), Rottnest island quokka selfie as popular as ever thanks to celeb snaps, in Dodds R. & Butler R. 2019, *Overtourism-Issues, realities and solutions*, De Gruyter Oldenbourg.

Ayeh, J., Leung, D., Au, N. & Law, R. (2012). 'Perceptions and strategies of hospitality and tourism practitioners on social media: an exploratory study'. In Fuchs, M., Ricci, F. & Cantoni, L., (Eds.), *Information and*

Communication Technologies in Tourism 2012, Proceedings of the International Conference in Helsingborg, Sweden, 25–27 January. Springer: Vienna, Austria.

Bansal H.S. and Voyer P.A. (2000). 'Word-of-mouth processes within a services purchase decision context', *Journal of Services Research,* 3 (2) 166-177.

Barcelona City Council, (2018). Barcelona tourism activity report 2017. http://www.barcelonaturisme.com/uploads/web/estadistiques/CAPSULA1_IA2018.pdf.

Belk, R.W. (2010). 'Sharing'. *Journal of Consumer Research,* 36(5), p715–734.

Boissevain, J.(1996). *Coping with Tourists: European reactions to mass tourism* (Vol. 1), Berghahn Books.

Bosselman, F. Peterson, C. & McCarthy, C. (1999). *Managing Tourism Growth: Issues and applications.* Island Press.

Bourliataux-Lajoinie, S. & Maubisson, L. (2017). 'Online review ratings and the impact on the number of visits to a tourist and cultural attraction: the case of the Loire Valley and TripAdvisor', *Journal of Internet Social Networking and Virtual Communities.* DOI: 10.5171/2018.915203.

Buhalis, D. & Law, R. (2008). 'Progress in information technology and tourism management: 20 years on and 10 years after the internet - The state of eTourism research', *Tourism Management,* 29(4), 609-623.

Casaló L.V., Flaviána C., Guinalíua M. & Ekinci Y. (2015). 'Do online hotel rating schemes influence booking behaviors?', *International Journal of Hospitality Management,* 49, 28-36.

Casaló, L.V., Flavián, C. & Guinalíu, M. (2011). 'Understanding the intention to follow the advice obtained in an online travel community', *Computers in Human Behavior,* 27, 622–633.

Chen, J.S. & Uysal, M.(2002). 'Market positioning analysis: A hybrid approach', *Annals of Tourism Research,* 29(4), 987–1003.

Chen, W., Wang, Y. & Yang, S. (2009, June). 'Efficient influence maximization in social networks'. In *Proceedings of the 15th ACM SIGKDD international conference on knowledge discovery and data mining* (pp. 199-208).

Chen, W., Yuan, Y. & Zhang, L. (2010). 'Scalable influence maximization in social networks under the linear threshold model'. In *2010 IEEE international conference on data mining* (pp. 88-97). IEEE.

Chon, K. S. (1991). 'Tourism destination image modification process: Marketing implications'. *Tourism Management,* 12(1), 68-72.

Davis, F. (1989). 'Perceived usefulness, perceived ease of use, and user acceptance of information technology', *MIS Quarterly*, 13 (3), 319–340.

Decrop, A. (2008). "Les paradoxes du consommateur postmoderne" in *Reflets et perspectives de la vie économique*, (2)47, 85-93.

Del Chiappa, G., Alarcon-del-Amo, M.-C. & Lorenzo-Romero, C. (2015), 'The influence of user-generated content on tourists' choices.', *Market-Trziste*, 27(2), 221-236.

Dickinger, A. (2011). 'The trustworthiness of online channels for experience- and goal-directed search tasks', *Journal of Travel Research*, 50(4), 378-391.

Dinhopl, A., & Gretzel, U. (2016). Selfie-taking as touristic looking. *Annals of Tourism Research*, 57, 126-139.

Filieri R., Alguezaui S. & McLeay F. (2015). 'Why do travelers trust TripAdvisor? Antecedents of trust towards consumer-generated media and its influence on recommendation adoption and word of mouth', *Tourism Management*, 51, 174-185.

Foley, A. & Fahy, J. (2004). Incongruity between expression and experience: The role of imagery in supporting the positioning of a tourism destination brand. *Journal of Brand Management*, 11(3), 209-217.

Fyall, A. & Garrod, B. (1998). 'Heritage tourism: at what price?'. *Managing Leisure*, 3(4), 213-228.

Gonzalez, V. M., Coromina, L. & Galí, N. (2018). 'Overtourism: residents' perceptions of tourism impact as an indicator of resident social carrying capacity-case study of a Spanish heritage town'. *Tourism Review*, 73(3), 277-296.

Gosling, S. D., Augustine, A. A., Vazire, S., Holtzman, N. & Gaddis, S. (2011). 'Manifestations of personality in online social networks: Self-reported Facebook-related behaviors and observable profile information'. *Cyberpsychology, Behavior, and Social Networking*, 14(9), 483-488.

Gretzel, U. (2019). The role of social media in creating and addressing overtourism. In Dodds, R. & Butler, R. (Eds.). *Overtourism: Issues, realities and solutions*, pp. 62-75. Berlin: De Gruyter.

Gretzel, U., Yoo, K. H. & Purifoy, M. (2007). 'Online travel review study: Role and impact of online travel reviews'. Austin, TX: Laboratory for Intelligent Systems in Tourism, Texas A&M University.

Gurviez P. & Korchia M. (2002). 'Proposition d'une échelle de mesure multidimensionnelle de la confiance dans la marque', *Recherche et Applications en Marketing*, 17 (3), 41-61.

Harrison, F. (2013). 'Digging deeper down into the empirical generalization of brand recall: Adding owned and earned media to paid-media touchpoints'. *Journal of Advertising Research*, 53(2), 181-185.

Hennig-Thurau, T., Gwinner, K. P., Walsh, G. & Gremler, D. D. (2004). 'Electronic word of mouth via consumer-opinion platforms: what motivates consumers to articulate themselves on the internet?', *Journal of Interactive Marketing*, 18(1), 38-52.

Hetzel, P. (2002). *Planète Conso : Marketing expérientiel et nouveaux univers de consommation*, Paris: Éditions d'Organisation.

Holbrook, M.B, (1999). *Consumer value: a framework for analysis and research*. Psychology Press.

Hossler M., Murat O. & Jouanne A. (2014) *Faire du marketing sur les réseaux sociaux: 12 modules pour construire sa stratégie social media*. Paris: Eyrolles.

Johnson, T.J. & Kaye, B.K. (2009). 'In blogs we trust? Deciphering credibility of components of the internet among politically interested internet users'. *Computers in Human Behavior*, 25, 175–182.

Kock, F. Zenker, S. Josiassen, A. Norfelt, A. & Wilke, R. (2018). 'The tourism dilemma: examining conflicts between tourists and residents', In 2018 Global Marketing Conference at Tokyo (pp. 635-636).

Koens, K., Postma, A. & Papp, B. (2018). Is overtourism overused? Understanding the impact of tourism in a city context. *Sustainability*, 10(12), 4384.

Korfiatis, N., García-Bariocanal, E. & Sánchez-Alonso, S. (2012). 'Evaluating content quality and helpfulness of online product reviews: The interplay of review helpfulness vs. Review content', *Electronic Commerce Research and Applications*, 11(3), 205-217.

Kuščer, K., & Mihalič, T. (2019). 'Residents' attitudes towards overtourism from the perspective of tourism impacts and cooperation—The case of ljubljana'. *Sustainability*, 11(6), 1823.

Larsen, J., Urry, J., & Axhausen, K. W. (2007). 'Networks and tourism: Mobile social life'. *Annals of Tourism Research*, 34(1), 244-262.

Lee HA, Law R & Murphy J. (2011). 'Helpful reviewers in TripAdvisor : an online communi-ty', *Journal of Travel & Tourism Marketing*, 28, 675-688.

Lovelock, C.H. &Wirtz, J.W. (2011). 'Services marketing: people, technology, strategy', *Journal of Services Marketing*, 18(5).

Lovett, M. J. & Staelin, R. (2016).' The role of paid, earned, and owned media in building entertainment brands: Reminding, informing, and enhancing enjoyment'. *Marketing Science, 35*(1), 142-157.

Mak, A.H.N. (2017). 'Online destination image: Comparing national tourism organisation's and tourists' perspectives', *Tourism Management*, 60, 280-297.

María-del-Mar, A. A., Borrajo-Millán, F. & Liu, Y. (2019). Are social media data pushing overtourism? The case of Barcelona and Chinese tourists. *Sustainability*, 11(12).

Martín, J.M.M., Martínez, J.M.G. & Fernández, J.A.S. (2018). 'An analysis of the factors behind the citizen's attitude of rejection towards tourism in a context of overtourism and economic dependence on this activity.' *Sustainability*, 10(8), 1-18.

Mauri, A. & Minazzi, R. (2013). 'Web reviews influence on expectations and purchasing in-tentions of hotel potential customers', *International Journal of Hospitality Management*, 34, 99–107.

Milano, C. (2017). 'Overtourism y Turismofobia: Tendencias globales y contextos locales.', Barcelona Ostelea School of Tourism & Hospitality.

Moon, J. H., Lee, E., Lee, J. A., Choi, T. R., & Sung, Y. (2016). 'The role of narcissism in self-promotion on Instagram'. *Personality and Individual Differences*, 101, 22-25.

Narangajavana, Y. Fiol, L.J.C. Tena, M.Á.M. Artola, R.M.R. & García J.S. (2017), 'The influence of social media in creating expectations An empirical study for a tourist destination', *Annals of Tourism Research*, 65, 60-70.

Nicholls, R. (2010) 'New directions for customer-to-customer interaction research;. *Journal of Services Marketing*, 24(1), 87–97.

Paquerot, M., Queffelec, A., Sueur., I. & Biot-Paquero G (2011). 'L'e-réputation ou le renforcement de la gouvernance par le marché de l'hôtellerie?', *Revue Management et Avenir*, 45, 280-296.

Plotkina D., Munzel A. & Pallud J. (2017). 'Don't let them fool you. Detecting fake online reviews', *Proceding of Association Française of Marketing Congress*, Tours, France.

Racherla P. & Friske W. (2012). 'Perceived "usefulness" of online consumer reviews: An exploratory investigation across three services categories', *Electronic Commerce Research and Applications*, 11(6), 548-559.

Schindler, R.M.. (1989). 'The excitement of getting a bargain: some hypotheses concerning the origins and effects of smart-shopper feelings', *Advances in Consumer Research*, 16, 447-453.

Sigala, M. (2017). 'Collaborative commerce in tourism: Implications for research and industry', *Current Issues in Tourism*, 20(4), 346–355.

Steen Jacobsen J.Kr & Mumar A.M, (2012). 'Tourist information search and destination choice in a digital age', *Tourism Management Perspectives*, 1, 39-47

Tuten T.L, Solomon M.R, (2016). *Social Media Marketing*, Thousand Oaks:Sage

Tyler, D., Guerrier, Y. & Robertson, M. (1998). *Managing Tourism in Cities: policy, process, and practice.* John Wiley & Sons.

Walmsley, A. (2017). 'Overtourism and underemployment: a modern labour market dilemma.', *Responsible Tourism in Destinations, Tackling Overtourism Conference*. Reykjavik, Iceland.

Xiang, Z. & Gretzel, U. (2010). 'Role of social media in online travel information search', *Tourism Management*, 31, 179–188.

Xu, Q. (2014). 'Should I trust him? The effects of reviewer profile characteristics on eWOM credibility', *Computers in Human Behavior*, 33, 136–144.

Yoo, J., Choi, S., Choi, M. & Rho, J. (2014). Why people use Twitter: social conformity and social value perspectives. *Online Information Review*.

11 Residents' perceptions of cruise tourism in an overcrowded city: The case of Venice

Giacomo Del Chiappa, Francesca Checchinato and Marcello Atzeni

The objectives of this chapter are to:

- ☐ Study and profile a quota sample of residents in Venice, a homeport in Italy characterized by overtourism, based on their perceptions and attitudes toward the development of cruise tourism.
- ☐ Introduce three clusters derived from the study findings, namely 'cautious', 'optimistic' and 'opposers' that significantly differ according to their economic reliance on cruise tourism, relatives' economic reliance on cruise activity, education level and geographical proximity to the cruise port area.
- ☐ Address both theoretical contributions and managerial implications, and give future research venues.

Keywords: Overtourism, cruise tourism, community-based tourism, Venice, cluster analysis, homeport Italy.

Introduction

Tourism is one of the most important industries in Europe: it represents 10% of the European Union GDP and 12 million people are employed in this sector (UNWTO, 2018). Due to its important contribution to the economy and its impact on the community, it affects the everyday life of residents, both in a positive and negative way. Within the industry, cruise tourism is the fastest growing segment of leisure tourism (Klein, 2011). In the last twenty years, the cruise sector has increased significantly, amounting to 24.7 million passengers in 2016 (CLIA, 2018) and employing 1,021,681 people around the world (BREA, 2017). Further, the cruise sector produces $57.9 billion in direct expenditures, thus creating a total economic output of $125.96 billion worldwide. In this scenario, academic research has devoted to analyze the residents' perceptions and attitudes toward cruise tourism development (i.e. Brida et al., 2011; Del Chiappa & Abbate, 2016). However, studies have mostly analyzed cruising destinations in the Caribbean, Arctic and the polar areas (Hritz & Cecil 2008; Diedrich 2010; Klein 2010; Stewart et al., 2013; Heeney, 2015; Stewart et al., 2015; Jordan & Vogt, 2017) and, recently, also in the Mediterranean area (Marušić et al., 2008; Brida et al., 2012; Peručić & Puh, 2012; Pulina et al., 2013; Del Chiappa & Abbate 2016; Del Chiappa et al., 2017; Del Chiappa, et al., 2018b; Del Chiappa, et al., 2018c), mainly focusing on port-of-call cruise destinations. Despite this, only few studies have been carried out on homeport cruise destinations so far (Brida & Zapata 2010), and very few studies exist on destinations where the number of tourists creates massive overcrowding.

This study was therefore carried out by surveying a quota sample of 354 individuals residing in Venice. Venice was selected as the research setting for this study for two main reasons. First, it is the second homeport in the Mediterranean area and one of the most famous tourism destinations worldwide, with around 24 million tourists a year. Second, it is considered to be affected by the so-called overtourism phenomenon (Seraphin et al., 2018). Anti-tourism movements have been growing in the last few years, voicing their concerns toward the continuous growth of the tourism phenomenon in the city, particularly toward cruise-related activities. This renders the research setting particularly interesting for the purposes of this study. Specifically, this paper aims to profile residents in Venice according to their perceptions towards the impacts of cruise tourism, and to ascertain whether there are significant differences among the clusters based on the socio-demographic traits of respondents.

Literature review

Considering the residents' perceptions, expectations and attitudes towards the impact of a prospective tourism development is pivotal for planning the future of any tourism destination (e.g. Sharpley, 2014), especially to achieve sustainable tourism development and to ensure the community members' support for tourism projects (Vargas-Sánchez et al., 2009). Most of the community-based studies are based on social exchange theory framework (Ap & Crompton, 1993) which links the costs and benefits for the local population with its willingness to support tourism. Indeed, residents will be more inclined to support tourism development when they perceive that the benefits are greater than the related costs. Recent academic studies mainly support this theory, revealing that local communities recognize both positive and negative consequences arising from tourism (Andereck et al., 2005) and their perceptions influence their support (or lack thereof) of tourism development.

In the last decades, community-based studies have sought to understand which factors can affect residents' perceptions and attitudes toward tourism development, and which of these are classifiable as extrinsic factors and intrinsic factors (Faulkner & Tideswell, 1997).

Extrinsic factors mainly refer to characteristics of a tourist destination (e.g. tourism seasonality, the stage of tourism development, tourist-guest ratio, etc.) (Fredline & Faulkner, 2000; Nyaupane et al., 2006) and they are likely correlated with the different life-cycle stage of the specific destination (Butler, 1980), where the attitude towards tourism development decreases over time with the growth of tourism flows (Papathanassis, 2017; Del Chiappa et al., 2018a). On the other hand, intrinsic factors refer to the residents' individual characteristics (e.g. socio-demographic characteristics), their environmental attitudes and beliefs, their economic reliance to tourism, and their perception of positive and negative impacts arising from the tourism development (e.g. Del Chiappa et al., 2018a). Social exchange theory proposes that when residents perceive that benefits from tourism are greater than costs, they are more willing to support tourism development (Ap, 1992).

Hence, recent studies of cruise tourism have moved to analyze the residents' perceptions and attitudes towards it (e.g. Hritz & Cecil, 2008; Brida et al., 2014; Del Chiappa et al., 2018c), thus expanding and deepening our understanding about the perceived impacts of cruise activity within the host community. Existing research highlights that residents usually show an overall slightly positive attitude towards cruise activity (e.g. Del Chiappa & Abbate, 2016; McCaughey et al., 2018), in particular when the cruise tourism highly affects the destination economy (McCaughey et al., 2018). But, at the same time, recent studies show that residents are not often so willing to support further cruise tourism development in their destination (Peručić & Puh, 2012), mainly showing concern for the issues of overcrowding and pollution (Klein, 2011), which are seen as ecological threats, damaging for example the marine ecosystem, and its real impact on local economy (Diedrich, 2010; Kerswill & Mair 2015). Further, when compared with different types of tourism, cruise tourism is not listed as the favorite type (e.g. Brida et al., 2012; Del Chiappa & Abbate, 2016; Del Chiappa et al., 2016; Del Chiappa et al., 2018b), possibly because it is strongly linked to the mass tourism concept of large groups of people visiting local areas (Diedrich, 2010).

Furthermore, existing studies have investigated the moderator effect of intrinsic factors (namely socio-demographic characteristics) on residents' perceptions and attitudes towards cruise activity, and provided results that are somehow contradictory. For example, in the case study of Messina, Del Chiappa and Abbate (2016) found significant differences among people of different ages, employment reliance, levels of education, geographical proximity to the tourism area, length of residency and contact frequency with tourists. On the contrary, in Valencia no significant differences were found in the residents' perception according to their different level of education or length of residence (Del Chiappa et al., 2018c). This finding is partially confirmed by Del Chiappa et al.'s study (2018b), where resident attitudes toward cruise tourism development significantly differ according to their prior experience with cruise vacations and their relatives' economic reliance on cruise activity, while no significant differences were found based on gender, age, employment status, economic reliance on cruise tourism, education level, length of residence and geographical proximity to the cruise port area.

In contrast, Jordan and Vogt (2017) found out that individuals with different levels of education had different perceptions of the impact of cruise tourism. Indeed, they underlined that residents with higher levels of education perceived different stressors: people with elementary school or high school education perceived stressors like "no tourists, no jobs, no money", whilst people with a university degree perceived stress elements like traffic congestion.

This suggests that community-based tourism studies in the context of cruise activity are highly site-specific and hardly generalizable (Del Chiappa et al., 2018c), making any effort to apply such type of investigation in other cruise tourism destinations particularly relevant, especially when homeports in tourism destinations affected by overtourism are considered.

Overtourism has been described as "the impact of tourism on a destination, or parts thereof, that excessively influences perceived quality of life of citizens and/or quality of visitors experiences in a negative way" (UNWTO, 2018). It is often correlated with negative factors, such as inappropriate behaviour of visitors, the touristification of the destination's city center, the marginalization of residents to residential areas (Koens et al., 2018), and the proliferation of new and non-institutionalized forms of tourist accommodation (Sarantakou & Terkenli, 2019; Smith et al., 2019). Recently, overtourism has received public attention, as a consequence of the anti-tourism and anti-cruise protests in some popular tourism destinations and specifically in some homeport destinations, and has become a central topic in tourism research (Goodwin, 2017; Papathanassis, 2017; Wall, 2020). Despite this, to the best of our knowledge, limited research has been devoted to analyze the residents' perception of cruise tourism impacts in homeport destinations affected by overtourism.

This study is therefore carried out to empirically investigate residents' perceptions towards cruise tourism and whether these attitudes are homogeneous, when considering a Mediterranean-based homeport cruise destination affected by overtourism (i.e. Venice, Italy). Specifically, in accordance with previous studies (i.e. Fredline & Faulkner 2000; Aguiló & Rosselló 2005; Brida et al., 2010; Presenza et al., 2013; Del Chiappa & Atzeni 2015) a quantitative method based on factor-cluster analysis was applied on a sample of 354 residents living in the city of Venice, in order to profile residents according to their perceptions of cruise activity and to ascertain whether significant differences exist between clusters of residents based on their sociodemographic characteristics.

Methodology

Venice is one of the main homeport cruise destinations of the Mediterranean area. In 2017, Venice attracted 1,446,635 cruise tourists (Venice Port Authority, 2019), who made up for 28.7% of total arrivals (5,034,882) in the city in the same year. In the last ten years, Venice has been increasingly suffering from tourist overcrowding, so that more and more residents are leaving the historic center of the city and moving to the hinterland.

In 2017, the Venice council launched the #enjoyrespectVenezia campaign providing guidelines on how to be a more responsible tourist, creating maps highlighting public toilets and picnic areas, and sharing a calendar illustrating tourist traffic forecasts. Moreover, during the same year, higher fines for inappropriate behavior were introduced. Due to the touristification of the city center – where the cost of living has become too high (Bertocchi & Visentin, 2019) and most of the activities are designed to boost tourism – residents are moving out of Venice to Terraferma (i.e. Venice's mainland, including all the towns close to Venice but not the islands). As a result, while 66,386 people lived in the city center in 2000, only 52,996 residents were in the city center in 2018 (Municipality of Venice, 2019). Many Venetians who leave the heart of the city end up renting out their apartments to tourists, through platforms such as Airbnb.

Furthermore, Venice is currently suffering from a major environmental issue: the ecosystem of the lagoon is becoming more and more precarious and the city is slowly sinking, as the streets and the buildings do not have proper foundations and they are gradually subsiding into the waters of the lagoon. For these reasons, Venetians are campaigning against cruise ships being allowed to dock in the heart of their city.

In this scenario, cruise tourism can be perceived by residents negatively, because its flows concentrate in an already crowded city. Therefore, because of the overtourism phenomenon and related tourismphobia, more criticism towards cruise impacts are expected, when compared to those reported in other homeport tourism destinations.

Even if cruise shipping generates large economic benefits for the city of Venice, costs are often emphasized more and discussed by residents. In 2012, a number of residents founded an activist group called Comitato No Grandi Navi ('No Big Ships Committee') to protest against cruise tourism and its environmental impact, further showing that cruise tourism suffers from a negative reputation among residents. One of the debated issues is the ships' route, considered too close to St Mark's square and to the cultural heart of the city. Plans to divert large cruise ships away from St Mark's basin and the Giudecca canal were drawn up, and different proposals to keep ships out of the city were made, but no consensus has been reached yet. In June 2019, a big ship crashed on the Marittima dock, igniting further protests. During the same year, the President of the North Adriatic Sea Port Authority launched the 'Cruise 2030 Call For Action' (Delepouve, 2019) in seven main European cruise ports, to outline common strategies to support the sustainability of cruise activities and to eliminate related externalities (e.g. waste, risks, overcrowding).

Despite the numerous newspapers and magazines' articles about cruise tourism and the anti-ships protest in Venice, to our knowledge no academic studies have yet investigated the residents' perspective towards the phenomenon.

These circumstances show that Venice is an interesting research setting to run any research aimed at investigating residents' views and attitude toward cruise tourism. For the purposes of the study, a survey was developed based on prior research on residents' perceptions and attitudes towards cruise tourism (e.g. Del Chiappa & Abbate, 2016), and items were slightly modified to suit the specific destination under investigation. The first part of the questionnaire invited respondents to assess their level of agreement with 28 statements related to economic, social and environmental impacts deriving from cruise tourism development (5-point Likert scale: 1= strongly disagree, 5 = strongly agree). The second part asked respondents to provide general socio-economic information (age, gender, level of education, employment reliance on cruise tourism, etc.). Data were collected in Venice in 2014, through face-to-face and online collection. At the end of the data collection, a total of 354 complete responses was obtained and used for the running statistics.

Results

The sample was mainly composed by females (55.6%) in the age bracket 26-35 years old (23.9%) or 36-56 years old (36.1%), with a high school degree (38.2%) or university degree (35.5%), working as an employee (45.6%) or self-employed (17.7%). Respondents have resided in Venice for more than 21 years (19.4% for 21-30 years; 37.0% for more than 31 years), close to the cruise port area (34.3% under 5 kilometres) and without any economic reliance on cruise activity (88.8%) (Table 11.1). Overall, the majority of respondents reported not having contact with tourists (58.8%) and specifically with cruise tourists (51.2%) in their daily life, and they declared that neither their job (88.8%) nor that of their relatives (84.7%) was related to tourism. Finally, 72.0% of residents interviewed have never had a cruise trip at the time of the data collection.

Table 11.1: Socio-demographic characteristics of the sample (%)

Gender	%	**Does your relative's job relate to tourism?**	%
Male	44.2	Yes	15.3
Female	55.8	No	84.7
Age	%	**How many years have you been residing in the city of Venice?**	%
18-25	15.2	< 5	5.4
26-35	23.9	6-10	0.6
36-56	36.1	11-20	7.5
over 56	23.6	21-30	19.4
no response	1.2	More than 31	37.0
Education	%	No response	30.1
No qualification	0.0	**Distance from home to cruise port**	%
Elementary school	1.2	Less than two	17.9
Secondary/high school	14.6	Between 3 and 5	16.4
Diploma/trade	38.2	Between 6 and 10	19.1
University degree	35.5	Between 11 and 20	11.0
Postgraduate degree	10.1	More than 21	0.9
No response	1.6	No response	34.7
Employment	%	**Have you ever had contacts with cruise tourists?**	%
Employee	45.6	Yes	48.8
Self-employed	17.7	No	51.2
Retired	14.7	**Have you ever been on a cruise trip?**	
Unemployed	2.7	Yes	28.0
Student	13.5	No	72.0
Other	5.7		
Does your job relate to tourism?	%		
Yes	11.2		
No	88.8		

Source: Authors' elaboration.

Overall, findings reveal that according to the residents, the negative effects of cruise tourism development significantly outweighed the positives ones, with responses scoring high or very high on items measuring negative social impacts (e.g. "Makes local entertainment facilities and public areas overcrowded"– M=4.01; "Produces significant levels of waste/

garbage" – M=4.02) and negative environmental impacts (e.g. "Increases air and marine pollution"– M=4.25; "Alters the ecosystem (sand erosion, flora and fauna are damaged, etc.) – M=4.11; "Increases air and marine pollution" – M=4.25). Despite this, respondents were reported scoring slightly positive on some of the statements used to measure positive economic impacts (e.g. "Increases job opportunities"– M=3.47; "Increases private investments and infrastructures" – M=3.38) (Table 11.2).

Table 11.2: Residents' views towards cruise tourism development: results of factor analysis

	Loadings	Eigenvalue	% of variance	% cumulated variance	Alpha
Positive socio-cultural / environmental impacts		13.144	41.074	41.074	0.937
Enhances the quality of life	0.568				
Allows to meet new people and to experience new culture	0.602				
Enhances the local offer of cultural entertainment activities and attractions	0.628				
Makes the best of this location's identity and authenticity	0.765				
Enhances the quality of restaurants, hotels and retail facilities	0.643				
Enhances social and cultural life for local people	0.773				
Incentivizes better infrastructures (roads, water, supply, etc.)	0.674				
Enhances the quality of public services	0.699				
Allows to preserve and to exploit the local cultural heritage	0.692				
Enhances urban and rural settings	0.781				
Improves the safety & security of the city	0.557				
Incentivizes the preservation of the environment	0.69				
Tourism is effectively managed in Venice	0.538				
Negative environmental impacts		2.505	7.829	48.903	0.909
Alters the ecosystem (sand erosion, flora & fauna are damaged, etc.)	0.786				
Increases air and marine pollution	0.828				
Serious damages to the city could occur	0.776				

Serious damages for the environment could occur	0.855				
Tourism should be regulated	0.516				
Negative social impacts		**1.691**	**5.283**	**54.186**	**0.795**
Forces me to change the way I manage my daily life	0.689				
Increases the cost of living	0.652				
Increases the number of minor crimes	0.631				
Makes local entertainment facilities and public area overcrowded	0.605				
Produces significant levels of waste/ garbage	0.602				
Cruise tourism influences the way I can manage my daily life	0.71				
Positive economic impacts		**1.358**	**4.244**	**58.43**	**0.787**
Increases job opportunities	0.719				
Increases public investments and infra-structures	0.583				
Increases private investments and infra-structures	0.688				
Increases the income of local people	0.612				

Source: Authors' elaboration.

For the purpose of this study, a factor cluster analysis was used. First, a factor analysis was applied and four factors emerged, explaining the 58.43% of total variance. The Bartlett test of sphericity (Chi squared=6649.429; sig=0.000) and the KMO index (KMO=0.938) indicated a good model acceptability (Hair et al., 2013). To test the reliability of factors, Chronbach's alpha was then calculated, and following Nunnally (1978), the reliability of the scales demonstrated high internal consistency of the constructs as Cronbach's alpha exceeded 0.70 (Table 11.2). The first factor ("Positive socio-cultural and environmental impacts" – 41.07% of total variance), is composed of items devoted to measure the positive effects that cruise tourism could generate, such as improving the quality of daily life for residents, enhancing local identity and authenticity and incentivizing sustainable environmentally practices. The second factor is named "Negative environmental impacts" (7.89% of total variance), and included items related to residents' concerns about the negative environmental impact that cruise activity generates: increasing air and marine pollution, altering the ecosystem and generating serious damage for the environment and the city as a whole. The third factor "Negative social impacts" (5.28% of total variance) consisted of items related to the negative social effects perceived by residents, such as the increase in the cost of living, the higher production of waste/garbage, the increase of the number of minor crimes and the higher level of overcrowding on public areas. Finally, the fourth factor labelled "Positive economic impacts" (4.24% of total variance) was related to the residents' perceptions of the positive impacts that cruise tourism could generate, such as increasing public and private investment, increasing job opportunities and income for local people.

A double-step cluster analysis was then performed to factor scores. First, a hierarchical cluster analysis was performed, (Ward method – Manhattan distances) and three cluster emerged. Afterwards, a non-hierarchical method was used (k-means method), allowing to identify three groups: 'Opposers', 'Cautious' and 'Optimistic' (Table 11.3).

Table 11.3: A comparative analysis of the level of agreement of different groups of residents (mean value)

	Cautious N=125	Optimistic N=76	Opposers N=153	Total N=354
Positive socio-cultural / environmental impacts				
Enhances the quality of life	2.47	3.17	1.33	2.13
Allows to meet new people and to experience new culture	3.08	3.43	1.71	2.56
Enhances the local offer of cultural entertainment activities and attractions	2.96	3.38	1.55	2.44
Makes the best of this location's identity and authenticity	2.67	3.29	1.45	2.28
Enhances the quality of restaurants, hotels and retail facilities	2.74	3.42	1.54	2.36
Enhances social and cultural life for local people	2.27	3.00	1.39	2.05
Incentivizes better infrastructures (roads, water, supply, etc)	2.88	3.05	1.64	2.38
Enhance the quality of public services	2.54	2.8	1.65	2.21
Allows to preserve and to exploit the local cultural heritage	2.62	3.26	1.6	2.32
Enhances urban and rural settings	2.3	3.04	1.37	2.06
Improves the safety and security of the city	2.94	3.16	2.00	2.58
Incentivizes the preservation of the environment	1.86	2.7	1.35	1.82
Tourism is effectively managed in Venice	2.66	3.00	1.66	2.3

Negative environmental impacts				
Alters the ecosystem (sand erosion, flora & fauna are damaged, etc)	4.43	2.41	4.69	4.11
Increases air and marine pollution	4.61	2.67	4.73	4.25
Serious damages to the city could occur	4.25	2.03	4.55	3.9
Serious damages for the environment could occur	4.54	2.09	4.69	4.08
Tourism should be regulated	4.69	3.46	4.73	4.44
Negative social impacts				
Forces me to change the way I manage my daily life	2.84	2.51	3.5	3.05
Increases the cost of living	3.66	2.75	4.06	3.64
Increase the number of minor crimes	3.23	2.21	3.07	2.94
Makes local entertainment facilities and public area overcrowded	4.34	2.95	4.27	4.01
Produces significant levels of waste/garbage	4.27	3.11	4.26	4.02
Cruise tourism influences the way I can manage my daily life	3.34	2.46	3.86	3.37
Positive economic impacts				
Increases job opportunities	4.20	4.14	2.54	3.47
Increases public investments and infrastructures	3.53	3.5	1.93	2.83
Increases private investments and infrastructures	3.91	3.84	2.73	3.38
Increases the income of local people	3.46	3.78	1.89	2.85

Source: Authors elaboration.

'Opposers' was the largest group (N= 153), composed of mostly females (56.7%) belonging to the age bracket 26-35 years (28.3%) or 36-56 years, employees (47.5%) or students (15.8%) with high school degree (40.0%) or university degree (33.3%), residing more than 31 years in Venice (44.2%) between 1 and 20 kilometers from the cruise port area. 15.8% have relatives involved in jobs related to the cruise tourism and 15.8% have a cruise tourism related business activity.

Finally, 72.0% of 'opposers' have never had a cruise trip in their life. Overall, they showed a very critical and negative view towards cruise tourism development and they did not perceive any positive impact from social (i.e. "Enhances social and cultural life for local people" – M=1.39; "Enhances the quality of life" – M=1.33), economic (i.e. "Increases public investments and infrastructures" – M=1.93; "Increases the income of local people" – M=1.89) or environmental standpoints (i.e. "Increases air and marine pollution" – M=4.73; "Serious damages for the environment could occur" – M=4.69). Further, they strongly believe that cruise activities make local facilities and public areas overcrowded (M=4.27), and they would like the regulation of tourism in the city (M=4.73).

'Cautious' (N=125) was the second largest cluster and has a slight majority of females (51.4%), aged between 36-56 years old (34.7%) or over 56 years old (25.0%), employed (38.6%) or self-employed (24.3%), with high school degree (37.5%), residing in Venice more than 31 years (33.3%), mostly between 6 and 10 kilometers from cruise port area. 21.1% of them is involved in jobs related to the cruise tourism (the highest percentage among clusters) and 31.0% of them declared that their relative's job is related to the cruise tourism (the highest percentage among clusters). 38% of respondents belonging to this group had a cruise trip in their life. The second cluster included residents that are worried about the negative impacts that cruise tourism could have on the environment (e.g." Increases air and marine pollution"– M=4.61; "Serious damages for the environment could occur" – M=4.54; "Incentivizes the preservation of the environment" – M=1.86) and on the daily quality of life ("Makes local entertainment facilities and public area overcrowded"– M=4.34; "Enhances social and cultural life for local people" – M=2.27). Further, Cautious respondents do not believe that cruise tourism development preserves and exploits the local identity and authenticity (M=2.62) and favoured more regulation of tourism in Venice (M=4.69). Despite this, they thought that cruise tourism development increases "job opportunities" (M=4.20), as well as private (M=3.91) and public (M=3.53) investments on infrastructures, and the income of local people (M=3.46).

The third cluster ('Optimistic': N=76) are mostly women (57.3%) aged between 36-56 years old (41.3%), employed (47.6%) with university degree (44.8%). The large majority of individuals within this cluster do not economically depend on cruise activity (2.2%) and 93.0% were reported to have relatives whose income is not cruise tourism-related. They have been living in Venice for more than 21 years (20.3% between 21 and 30 years; 32.2% more than 31 years), reside close to the cruise port area (24.5% less than 2 km, the highest percentage among clusters) and have never gone on a cruise trip (78.0%). This group included respondents scoring slightly positive on items devoted to measure the positive socio-cultural and environmental impacts of cruise activity, whilst they are concerned about the negative environmental and social impacts arising from cruise activities.

For example, they agree that cruise tourism allows residents to meet new people and to experience new culture (M=3.43) and that it enhances the local offer of cultural entertainment activities and attractions (M=3.38) and the quality of restaurants, hotels and retail facilities (M=3.42). At the same time, they do not report concern about cruises' negative impacts on the ecosystem (M=2.41) or on the environment (M=2.09) arising from the cruise tourism. Despite this, they express a neutral position regarding the way cruise tourism development is managed in the city (M=3.0), and slightly agree with more regulation of the tourism phenomenon in the city (M=3.46).

Finally, a series of statistical tests (chi-squared and ANOVA) were run to test the existence of differences between clusters based on socio-economic and demographic characteristics of respondents (age, gender, employment status, education level, economic reliance on cruise activity, relatives' economic reliance on cruise activity, geographical proximity to cruise port area, length of residence, contact with tourists in everyday life and prior experience with cruise vacation). Findings reveal that differences exist based on the respondents' economic reliance on cruise tourism (X^2=20.85, p=0.00), their relatives' economic reliance on cruise activity (X^2=20.96, p=0.00), their education level (X^2=34.05, p=0.00) and geographical proximity to cruise port area (X^2=35.10, p=0.00). In contrast, no significant differences exist based on the respondents' gender (X^2=0.74 p=0.69), age (X^2=6.60, p=0.58), employment status (X^2=8.43, p=0.59), length of residence (X^2=14.29, p=0.16), contact with tourists in everyday life (X^2=14.19, p=0.165) and prior experience with cruise vacation (X^2=5.66, p=0.006).

Conclusion

This study was carried out in order to deepen the scientific debate on the residents' views of cruise tourism development in tourism destinations, specifically in homeport cruise destinations affected by overtourism, which represents an under-investigated research area. This study presented and discussed the findings of an empirical study carried out in Venice (Italy), one of the most famous overcrowded homeport destinations in the world. A factor-cluster analysis was applied to a sample of 354 residents.

Overall, our findings revealed that residents in Venice believe that the negative effects of cruise tourism development significantly outweighed the positive ones. On the whole, respondents reported critical views towards cruise impacts more when compared to those surveyed in port-of-call tourism destinations (Del Chiappa et al., 2016; Del Chiappa, et al., 2016; Del Chiappa & Abbate, 2016). This occurred despite the fact that residents in homeports were expected to express more positive views towards the cruise tourism impacts, since existing studies have proved that the economic impact of cruise activity is higher when homeports are considered (e.g. Brida & Zapata, 2010). This suggests that residents are strongly concerned about the significant contribution that cruise activity has in generating overtourism and related negative externalities. Furthermore, the factor-cluster analysis identified three segments ('cautious', 'opposers' and 'optimistic'), with significant differences based on specific socio-economic and demographic characteristics of respondents (i.e. economic reliance on cruise tourism, economic reliance on cruise activity of their relatives, level of education and geographical proximity to cruise port area). No significant differences were found based on other socio-demographic characteristics (i.e. gender, age, length of residency, prior onboard experience).

Our findings provide some contradictory results when compared with previous studies. For example, they confirm prior studies reporting educational level as a moderator of residents' perception (Jordan & Vogt, 2017). At the same time, they contradict prior studies reporting gender being a moderator factor of residents' views towards cruise activity (e.g. Brida et al., 2011; Nunkoo & Gursoy, 2012), and those reporting that residents perceive the positive effects of cruise activity when living close to the port area (Belisle & Hoy, 1980). Further, they contradict prior studies (e.g. Del Chiappa et al., 2018b) reporting that residents with prior cruise vacation experience express significantly different perceptions and attitudes toward cruise tourism, compared to their counterparts. This could be explained by the fact that residents in Venice are particularly involved and interested in significant problems associated with overtourism that affect, above all, the people living close to the center and the port area.

From a managerial perspective, the study provides relevant insights for policymakers and destination marketers, who should consider the perceptions of residents in an appropriate manner, when trying to determine whether the (perceived) carrying capacity has been reached and whether regulations activities need to be implemented to protect the interest of the local community and to avoid undermining the quality of the host-guest interactions. Furthermore, our findings suggest that policy makers and destination marketers should involve the local community in tourism planning better, and/or they could create activities and projects aiming at developing cruise tourism in a more sustainable way (Del Chiappa, 2012; Papathanassis, 2017), thus avoiding the residents' main perception of negative impacts of cruise activity, over the positive ones.

The fact that clusters differed based on certain socio-demographic characteristics does suggest that policy makers and destination marketers should take into account these variables when they plan internal marketing and communication activities. For instance, our findings reveal that significant differences among groups exist based on economic reliance on cruise activity and relatives' economic reliance on cruise activity, with residents or relatives who do not have an economic reliance on cruise activity being more critical than their counterparts. In this vein, it could be useful to deliver messages that focus on the positive impacts of cruise tourism on the local community, and to improve plans and activities devoted to involve residents in tourism planning and in tourism business activities.

While this study contributes to the literature and provides implications for practitioners, it is not free of limitations. Specifically, it has to be considered highly site specific and based on the use of a quota sample: hence, its findings can be hardly generalized, confirming the highly site-specific nature of community-based studies (Sharpley, 2014; Almeida et al., 2015; Del Chiappa et al., 2016; Del Chiappa et al., 2018c). For this reason, further research is needed to combine a broader set of characteristics (both intrinsic and extrinsic) that may moderate the residents' views towards cruise tourism, as well as to replicate the study in other homeport cruise destinations to cross-compare findings, and verify whether findings can be generalized.

References

Aguiló, E. & Roselló, J. (2005) 'Host community perceptions. A cluster analysis' *Annals of Tourism Research*, 32(4), 925-941.

Andereck, K. L., Valentine, K. M., Knopf, R. C. & Vogt, C. A. (2005) 'Residents' perceptions of community tourism impacts', *Annals of Tourism Research*, 32(4), 1056-1076.

Ap, J. (1992) 'Residents' perceptions on tourism impacts' *Annals of Tourism Research*, 19(4), 665-690.

Ap, J. & Crompton, J. L. (1993) 'Residents' strategies for responding to tourism impacts', *Journal of Travel Research*, 32(1), 47-50.

Belisle, F. J. & Hoy, D. R. (1980) 'The perceived impact of tourism by residents: A case study in Santa Maria, Columbia', *Annals of Tourism Research*, 12(1), 83–101.

Bertocchi, D. & Visentin, F. (2019) 'The Overwhelmed City": Physical and social over-capacities of global tourism in Venice', *Sustainability*, 11(6937). doi:10.3390/su11246937..

BREA (2017) 'The Global Economic Contribution of Cruise Tourism 2016'. http://cruising.org/docs/defaultsource/research/global_economic_contribution _2016.pdf?sfvrsn=2 (Accessed on 17.09.2018).

Brida, J. G. Del Chiappa, G., Meleddu, G. & Pulina, M. (2014) 'A comparison of residents' perceptions in two cruise ports in the Mediterranean Sea.' *International Journal of Tourism Research*, 16(2), 180-190.

Brida, J. G., Del Chiappa, G., Meleddu, M. & Pulina, M. (2012) 'The perceptions of an island community towards cruise tourism: A factor analysis', *Tourism*, 60(1), 29-42.

Brida, J.G., Riaño, E. & Aguirre S.Z. (2011) 'Residents' attitudes and perceptions towards cruise tourism development: A case study of Cartagena de Indias (Colombia)', *Tourism and Hospitality Research*, 11(3), 181-196. https://doi.org/10.1177/1467358411415153.

Brida, J. G. & Zapata, S. (2010) 'Economic impacts of cruise tourism: The case of Costa Rica', *Anatolia*, 21(2), 322-338.

Butler, R. W. (1980) 'The concept of a tourism area cycle of evolution: Implications for management resources', *The Canadian Geographer*, 24(1), 5-16.

Cruise Lines International Association (CLIA) (2018) '2018 Cruise Industry Outlook'. https://cruising.org/docs/default-source/research/clia-2018-state-of-the-industry.pdf (Accessed on 17.09.2018).

Del Chiappa, G. (2012) 'Community integration: A case study of Costa Smeralda, Italy'. In: Fayos-Sola, E., Silva, J., & Jafari, J. (Eds.). (2012), *Knowledge management in tourism: Policy and governance applications bridging tourism theory and practice*. Emerald: Bingley, 243–263.

Del Chiappa, G. & Abbate, T. (2016) 'Island cruise tourism development: a resident's perspective in the context of Italy', *Current Issues in Tourism*, 19(13), 1372–1385. https://doi.org/10.1080/13683500.2013.854751.

Del Chiappa, G. & Atzeni, M. (2015) 'Collaborative policy making: A community-based perspective in the context of Sardinia's Maddalena Archipelago, Italy'. In: Gursoy, E., Saayman, D., & Sotiriadis, M. (Eds.). *Collaboration in Tourism Businesses and Destinations: A handbook*. Emerald: Bingley, 57–74.

Del Chiappa, G., Atzeni, M. & Ghasemi, V. (2018a) 'Community-based collaborative tourism planning in islands: A cluster analysis in the context of Costa Smeralda', *Journal of Destination Marketing and Management*, 8, 41-48.

Del Chiappa, G., Atzeni, M., J. Pung & M. Risitano (2018b) 'Residents' views on cruise tourism in Naples Profiles and insights from a Mediterranean home-port destination', *European Journal of Tourism Research*, 23, 71-85.

Del Chiappa, G., Lorenzo-Romero, C. & Gallarza, M. (2018c) 'Host community perceptions of cruise tourism in a homeport: A cluster analysis', *Journal of Destination Marketing and Management*, 7,170–181. https://doi.org/10.1016/j.jdmm.2016.08.011.

Del Chiappa, G., Melis, G. & Atzeni, M. (2017) 'Turismo crocieristico e percezioni della popolazione residente: i risultati di un'analisi cluster', *Sinergie Italian Journal of Management*, 35(104), 135-154.

Delepouve H. (2019) The preservation of Venice and of its citizens can no longer wait. http://www.aivp.org/en/2019/12/19/the-preservation-of-venice-and-of-its-citizens-can-no-longer-wait/ (Accessed on 10.05.2019).

Diedrich, A. (2010) 'Cruise ship tourism in Belize : The implications of developing cruise ship tourism in an ecotourism destination', *Ocean and Coastal Management*, 53(5-6), 234-244. https://doi.org/10.1016/j.ocecoaman.2010.04.003.

Faulkner, B. & Tideswell, C. (1997) 'A framework for monitoring community impacts of tourism', *Journal of Sustainable Tourism*, 5(1), 3-28. https://doi.org/10.1080/09669589708667273.

Fredline, E. & Faulkner, B. (2000) 'Host community reactions: A cluster analysis', *Annals of Tourism Research*, 27(3), 763– 784.

Goodwin, H. (2017) 'The challenge of overtourism. Responsible Tourism Partnership'.https://haroldgoodwin.info/pubs/RTP'WP4Overtourism01'2017.pdf.

Hair, J. F. Black, W. C. Babin, B. J. & Anderson, R. E. (2013) *Multivariate Data Analysis*, 7th edn, Edinburgh Gate, Harlow, Essex: Pearson Education.

Heeney, E. (2015) 'Connection and understanding: the basis of a positive mutual gaze between residents of a small island developing state and a community of multinational ocean cruisers', *World Leisure Journal*, 57(2), 118-130.

Hritz, N. & Cecil, A. (2008) 'Investigating the sustainability of cruise tourism: A case study of Key West', *Journal of Sustainable Tourism*, 16(2), 168-181.

Koens, K. Postma, A. & Papp, B. (2018) 'Is overtourism overused? Understanding the impact of tourism in a city context', *Sustainability*, 10(12), 4384.

Kerswill, M. & Mair, H. (2015) 'Big ships, small towns: Understanding cruise port development in Falmouth, Jamaica', *Tourism in Marine Environments*, 10(3-4), 189-199. doi:10.3727/ 154427315X14181438892766.

Klein, R. A. (2010) 'Cruises and bruises: safety, security and social issues on polar cruises' in M. Luck, P.T. Maher and E. Steward (ed.), *Cruise Tourism in Polar Regions: Promoting Environmental and Social Sustainability*, London: Earthscan Ltd, pp. 54–74.

Klein, R. A. (2011). 'Responsible cruise tourism: Issues of cruise tourism and sustainability', *Journal of Hospitality and Tourism Management*, 18(1), 107-116.

Koens, K., Postma, A. & Papp, B. (2018) 'Is overtourism overused? Understanding the impact of tourism in a city context', *Sustainability*, 10(12), 4384.

Jordan, E. J. & Vogt, C. A. (2017) 'Residents' perceptions of stress related to cruise tourism development', *Tourism Planning and Development*, 14(4), 527-547. https://doi.org/10.1080/21568316.2017.1287123.

Marušić, A., Horak, S. & Tomljenović, R. (2008) 'The socio-economic impacts of cruise tourism: A case study of Croatian destinations', *Tourism in Marine Environments*, 5(2-3), 131-144.

McCaughey, R., Mao, I. & Dowling, R. (2018) 'Residents' perceptions towards cruise tourism development: the case of Esperance, Western Australia', *Tourism Recreation Research*, 43(3), 403-408.

Municipality of Venice (2019) Demographic statistics. https://www.comune.venezia.it/it/statistica (Accessed on 10.06.2019).

Nyaupane, G. P., Morais, D. B. & Dowler, L. (2006) 'The role of community involvement and number/type of visitors on tourism impacts: A controlled comparison of Annapurna, Nepal and Northwest Yunnan, China', *Tourism Management*, 27(6), 1373-1385.

Nunnally, J. (1978). *Psychometric Methods*. New York: McGraw-Hill.

Papathanassis, A. (2017) 'Overtourism and anti-tourist sentiment: An exploratory analysis and discussion'. *Ovidius University Annals, Economic Sciences Series*, 17(2), 288-293.

Peručić, D. & Puh, B. (2012) 'Attitudes of citizens of Dubrovnik towards the impact of cruise tourism on Dubrovnik', *Tourism and Hospitality Management*, 18(2), 213-228.

Presenza, A., Del Chiappa, G. & Sheehan, L. (2013) 'Residents' engagement and local tourism governance in maturing beach destinations: Evidence from an Italian case study', *Journal of Destination Marketing and Management*, 2(1), 22–30.

Pulina, M., Meleddu, M. & Del Chiappa, G. (2013) 'Residents choice probability and tourism development', *Tourism Management Perspectives*, 5, 57-67.

Sarantakou, E. & Terkenli, T. S. (2019) 'Non-institutionalized forms of tourism accommodation and overtourism impacts on the landscape: the case of Santorini, Greece', *Tourism Planning and Development*, 16(4), 411-433.

Seraphin, H., Sheeran, P. & Pilato, M. (2018) 'Overtourism and the fall of Venice as a destination', *Journal of Destination Marketing and Management*, 9, 374-376.https://doi.org/10.1016/j.jdmm.2018.01.011.

Sharpley, R. (2014) 'Host perceptions of tourism: A review of the research', *Tourism Management*, 42, 37-49.

Smith, M. K., Sziva, I. P. & Olt, G. (2019) 'Overtourism and resident resistance in budapest', *Tourism Planning and Development*, 16(4), 376-392.

Stewart, E. J., Dawson, S. E. L., Johnston, M. E., Pearce, T. & Lemelin, H. (2013) 'Local-level responses to sea ice change and cruise tourism in Arctic Canada's Northwest Passage'. *Polar Geography*, 36, 142-162.

Stewart, E., Dawson, J. & Johnston, M. (2015) 'Risks and opportunities associated with change in the cruise tourism sector: community perspectives from Arctic Canada', *The Polar Journal*, 5(2), 403-427.

UNWTO (2018) *Report on Tourism and Culture Synergies*. Madrid: UNWTO. https://www.e-unwto.org/doi/pdf/10.18111/9789284419876 (Accessed on 17.09.2018).

Vargas-Sánchez, A., Plaza-Mejía, M. & Porras-Bueno, N. (2009) 'Understanding residents' attitudes toward the development of industrial tourism in a former mining community', *Journal of Travel Research*, 47(3), 373-387.

Venice Port authority (2019) The port in figures. https://www.port.venice.it/en/the-port-in-figures.html (Accessed on 10.06.2019).

Wall, G. (2020) 'From carrying capacity to overtourism: a perspective article', *Tourism Review*, 1(1), https://doi.org/10.1108/TR-08-2019-0356.

12 Place attachment and residents' perceptions of tourism development in small town destinations

Carla Silva, Cláudia Seabra, José Luís Abrantes, Manuel Reis and Andreia Pereira

The objectives of this chapter are to:

- ☐ Analyse the relationship between place attachment and tourism impact the way it is perceived by small town destinations residents.
- ☐ Deepen the research on place attachment and on tourism impacts.
- ☐ Bring insights to small town destinations management and marketing.

Keywords: place attachment, tourism impacts, urban tourism, sustainability.

Acknowledgments: This work was funded by national funds through FCT – the Portuguese Foundation for Science and Technology (UID/ECO/00124/2013 and Social Sciences DataLab, Project 22209), POR Lisboa (LISBOA-01-0145-FEDER-007722 and Social Sciences DataLab, Project 22209), POR Norte (Social Sciences DataLab, Project 22209) and under the projects UIDB/05583/2020 and UIDB/04084/2020. Furthermore, we would like to thank CEGOT – Geography and Spatial Planning Research Centre, and Research Centre in Digital Services (CISeD), the Polytechnic of Viseu and the Faculty of Arts & Humanities of the University of Coimbra for their support. Also, we would like to thank the Instituto Politécnico de Viseu.

Introduction

Towns and cities have always been places where tourism and leisure experiences are constantly produced and consumed and a source of special fascination for visitors and tourists (Hall & Page, 2014). The development of tourism generates different degrees of impact on destination places, environments, and on the local population. Understanding resident perception of those impacts is crucial to the successful and sustainable development of tourism (Šegota, Mihalič & Kuščer, 2017). Residents are the most important destination stakeholders and, because of that, they should participate in the planning of sustainable tourism development in order to control the impacts of tourism on the places where they live (Lawton & Weaver, 2015; Garrod, Fyall, Leask & Reid, 2012). As a consequence, it is important to establish awareness of host communities' experiences, perceptions, and attitudes towards sustainable tourism development, particularly in locations with a small population base (Thompson-Fawcett & McGregor, 2011).

Once a community becomes a tourist destination, the lives of its residents become affected. Even though most of the studies conducted on tourism impacts focused on economic, social, and environmental wellbeing, the central question is still the need to understand how community residents perceive the impacts of tourism (Kim, Uysal, & Sirgy, 2013).

Another crucial concept regarding sustainable tourism development is place attachment, commonly defined as the emotional bond between an individual and a specific place (Manzo, 2003). However, affection, emotions, and feelings are not the only concepts of place attachment. Cognition and practice are important as well (Low & Altman, 1992; Vorkinn & Riese, 2001). People may feel attached to a place because of emotional and social ties, but also because of the physical aspects of the place (Hidalgo & Hernández, 2001; Lewicka, 2011). Places are above all social constructions and include physical, social, and psychological connotations that help build attachment (Knez, 2014).

Crucial to the role of place attachment in place-related studies is the ability to explain perceptions and behaviors in people's interactions with places as a result of affection. And although place attachment is a critical factor shaping residents' attitudes and perceptions toward tourism development, investigation relating those two dimensions, and particularly research focusing on small town destinations, is still scarce (Stylidis, 2018).

When residents develop deeper bonds with their nearby settings, they are more willing to be involved in the development of their regions, thus, more sensible to the effects of tourism. Within this context, this paper aims to analyze the relationship between place attachment and tourism impacts perceived by the residents of small town destinations, considering that there might be a positive relationship between residents' level of place attachment and their perceptions of the impacts that tourism may have on that place.

Literature review

Residents' perceptions of tourism impacts

Tourism influences residents' values, behaviors, lifestyles, and quality of life (Jaafar, Rasoolimanesh & Ismail, 2017; Rasoolimanesh, Badarulzaman, Abdulah & Behrang 2019; Gannon, Rasoolimanesh & Taheri, 2020) and affects positively and negatively their daily life. The quality of tourism experiences depends largely on the residents' quality of life. Therefore, "understanding resident perceptions and responses is fundamental to the successful and sustainable development of tourism" (Sharpley, 2014).

In general, tourism impacts are categorized positively and negatively (Jaafar, Noor, & Rasoolimanesh, 2015) into economic, environmental and socio-cultural impacts (Mathieson & Wall, 1982; Caneday & Zeiger, 1991; Gursoy & Rutherford, 2004; George, 2010). Recent studies on tourism impacts emphasis the way those effects are perceived by residents. There are several studies in tourism literature that focus on how tourism impacts are perceived by residents (e.g. Wang & Pfister, 2008; Nunkoo & Ramkissoon, 2010; Látková & Vogt, 2012; Kim, Uysal & Sirgy, 2013; Rasoolimanesh, Jaafar, Kock & Ramayah 2015; Rasoolimanesh, Ali & Jafaar, 2018).

In terms of socio-cultural effects, tourism can preserve the local cultural identity, enhance local culture and residents' pride (Rasoolimanesh et al., 2018), increase recreation activities and opportunities for residents (Gannon et al., 2020), and improve quality of life (Almeida-García, Peláez-Fernández, Balbuena-Vázquez & Cortés-Macias, 2016). But tourism development can also disrupt the normal routines of local residents (Lundberg, 2017) causing negative impacts on local communities such as overcrowding, traffic congestion, higher prices (Monterrubio, 2016), and crime (Rasoolimanesh et al., 2017, 2018).

As far as economic impacts are concerned, tourism creates more and new jobs for residents (Almeida-García et al., 2016; Rasoolimanesh et al., 2017, 2018), attract more investment (Rasoolimanesh et al., 2018; Gannon et al., 2020), increases the residents' standard of living (Gannon et al., 2020; Rasoolimanesh et al., 2017, 2018) , and provides higher income and more infrastructure and public facilities (Rasoolimanesh et al., 2017, 2018; Gannon et al., 2020). However, tourism also contributes to increase the cost of living (Rasoolimanesh et al., 2017), the price of some goods and services, and increases the price of propriety and land (Lundberg, 2017).

Finally, in terms of environmental effects, tourism can preserve the natural environment, helps preserve the historic buildings, and helps spread the image of a given place (Šegota et al., 2017; Gannon et al., 2020). On the other hand, it also creates noise and pollution (Rasoolimanesh et al., 2018) and increases litter in the community (Lundberg, 2017), which affects the residents' quality of life. Resident' perceptions and support for tourism development are influenced by diverse yet interconnected factors that include residents' place attachment (Gannon et al., 2020).

Place attachment

In recent years, research addressing the way people relate to places has become more prominent (Moore & Scott, 2003). Place attachment is a concept that derived from environmental psychology (Low & Altman, 1992) and has been applied to tourism. It describes an emotional link between people and places (Shumaker & Taylor, 1983).

Due its nature, place attachment is considered a multidimensional concept (Halpenny, 2010) comprising mostly two dimensions (Sharpley, 2014): *place identity* (Kyle, Absher & Graefe 2003, 2004; Yuksel, Yuksel & Bilim, 2010; Tsai, 2012; Chen, Dwyer & Firth, 2014a, 2014b, Woosnam et al., 2018) and *place dependence* (Williams, Patterson, Roggenbuck & Watson 1992; Kyle et al., 2003; Chen et al., 2014b; Yuksel et al., 2010; Tsai, 2012; Lundberg, 2017). These dimensions are distinct, because place identity is considered an emotional attachment and place dependence is regarded as a functional connection. This framework has been widely accepted in the tourism literature (Chen & Dwyer, 2018).

But there are several other studies that consider other dimensions, like: *affective attachment* (Chen et al., 2014a; Yuksel et al., 2010; Tsai, 2012; Ramkissoon, Weiler & Smith, 2012; Ratcliffe & Korpela, 2016), *social bonding* (Hidalgo & Hernández, 2001; Kyle, Bricker, Graefe & Wickham., 2004; Chen et al., 2014a; Ramkissoon & Mavondo, 2015; Ratcliffe & Korpela, 2016) *family/friend bonding* (Raymond, Brown & Weber, 2010), and *place memory* (Lewicka, 2011; Chen et al., 2014a; Chen & Dwyer, 2018).

Place attachment is often associated with other constructs such as residents' place satisfaction (Chen et al., 2014b, 2015) and personal involvement (Hou, Lin & Morais, 2005; Gross & Brown, 2008; Lewicka, 2011;

Prayag & Ryan, 2012; Tsai, 2012). Nonetheless, few studies relate residents' place attachment with residents' perceptions of tourism effects. Thus, and based on literature (Hidalgo & Hernandez, 2001) this study defines place attachment as an affective bond between people and particular places, suggesting that place attachment influences the residents' perceptions of tourism impacts.

Methodology

The research setting used a survey approach applied in Portugal, a European country, to residents of small towns in its Central Region. The empirical collection was conducted from October to December 2018.

Measurements and data collection

The survey instrument was built based on scales previously established in the tourism literature: i) to measure place attachment, we used the scales designed by Proshansky, Fabian and Kaminoff (1983), Gu and Ryan (2008); Kyle and colleagues (2004); and ii) for tourism impacts, the scales used were those built by Mathieson and Wall (1982), Krippendorf (1987), Liu et al., (1987), Crouch and Ritchie (1999), Starr (2002), Mason (2003) and Nepal & Chipeniuk (2005).

The survey instrument comprised two sections. The first one captured, using a 50 item scale, the residents' perceptions of tourism impacts considering their positive, negative, cultural, social, economic, and environmental effects. The second section measured residents' place attachment focusing on place identity and place dependence, using 16 items. In both sections, a 5-point rating scale was used. Respondents could express their level of agreement to a certain statement in 5 points where 1 refers to 'strongly disagree' and 5 to 'strongly agree'.

To assess the adequacy of the scales, the final instrument was discussed with several researchers and destination managers. In order to refine the instrument, the original scales were translated into Portuguese and subsequently the instrument was translated back into English. Afterwards, a pre-test was applied to 30 undergraduate students in order to test the scales' reliability through Cronbach's alpha (Cronbach, 1951). The results of this pre-test were used to make additional and final refinements to the questionnaire.

The collection of the data from residents was carried out using a convenience sample, social media and online platforms, but most of the data, about 75%, was collected in loco across different small towns in the Centre of Portugal. To be part of the sample and be allowed to fill in the questionnaire, the respondents had to be locals and to live in the town selected for at least three years. Residents were approached randomly, and the questionnaires were self-administrated to ensure data unbiasedness. 329 questionnaires were applied and 300 were considered valid.

Sample profile

The main elements of the sample profile are presented in Table 12.1. The sample had marginally more men (50.7%) than women (49.3%). Just under half were single (46.7%). 66.2% of them graduated from high school or dropped out before graduating from high school. The most common occupations were student (27.6%) or Sales/Administrative/Factory worker (22.2%). Most of them have a monthly individual income up to 1000 Euros (73.6%) or ranging between 1001 and 2000 Euros (23%).

Table 12.1: Sample Profile

Gender	Female	148	49.3%
	Male	152	50.7%
Age	Up to 25 years old	104	34.7%
	26-35 years old	80	26.7%
	36-45 years old	42	14.0%
	46-55 years old	42	14.0%
	56-65 years old	18	6.0%
	Over 65 years old	14	4.7%
Marital Status	Single	140	46.7%
	Married	135	45.0%
	Divorced	13	4.3%
	Widowed	11	3.7%
Educational level	University	101	33.8%
	High School or less	198	66.2%
Occupation	Sales/Administrative/Factory workers	66	22.2%
	Self-employed/Freelancer	31	10.4%
	Middle and Senior Management	8	2.7%
	Unemployed/Retired	38	12.8%
	Student	82	27.6%
	Public agent	47	15.8%
	Other	21	7.1%
	Work in tourism	34	11.3%
Income	Below 1000 Euros	192	73.6%
	Between 1001 and 2000 Euros	60	23.0%
	Over 2001 Euros	9	7.1%

Source: Authors' elaboration.

Results

This study presents the results of a quantitative survey that included 350 small town residents in Portugal.

The results are consistent with the scales used to measure the impacts of tourism development. Correlations between the scales' items range from -0.196 to 0.499. These values demonstrate that the items can be included in the Tourism Impacts dimensions. For place attachment, items correlations range between 0.424 and 0.760. The relational structure of tourism impacts and place attachment dimensions was analyzed using exploratory factor analysis over the correlation matrix, and factor extraction was achieved using principal component analysis and varimax rotation.

It was essential to identify several alternative solutions in order to guarantee the best factor structure. Additionally, the number of factors obtained in the analysis also took into account prior researches (Hair, Black, Babin, & Anderson, 2010)

To obtain the factor that measures tourism impacts, several factor analyzes were carried out and some variables were removed because they had factor loadings greater than 0.5 in more than one factor. The final model showed a Kaiser-Meyer-Olkin (KMO) Measure of Sampling Adequacy satisfactory at 0.894 (Sharma, 1996). The results obtained from Bartelett's Sphericity test show that the variables are significantly correlated (p-value = 0.000). They also indicated that the use of factor analysis is adequate (Sharma, 1996). Data showed statistical validity and a varimax rotation was performed. Factors were expected to be unrelated (West, 1991). Three factors explained 60.2% of the total variance in the sample. Table 12.2 shows those three factors and their respective factor loadings, variable communalities, factor explained variance and Cronbach's alphas (Cronbach, 1951).

Another factor analysis was conducted with Place Attachment items using the same procedure as described above. In this case, the KMO value obtained was 0.964 and the Bartelett's sphericity test showed that variables are significantly correlated (p-value = 0.000), showing that the use of the factor analysis procedure is suitable, and one factor emerged explaining on its own 64.8% of the total variance (Table 12.3).

Table 12.2: Factor loadings, variance and Cronbach's Alpha of cognitive destination image dimensions of small town destination residents, related to impacts of tourism development

	Factor loadings	Communalities	Eigen-values	% total variance	Cronbach's Alpha
Positive socio-cultural impacts					
Increases interest in other cultures	0.631	0.407	6.242	25.9	0.886
Creates opportunities for using recreation facilities	0.720	0.541			
Develops a wide variety of cultural activities	0.760	0.578			
Improves tolerance towards other cultures	0.663	0.472			
Provides social interaction	0.690	0.486			
Represents a worthy free time activity	0.729	0.538			
Provides an incentive for the preservation/ restoration of local historical heritage	0.755	0.603			
Breathes life and vitality into the community and the destination	0.773	0.618			
Improves the image of the city	0.768	0.620			
Negative social-cultural impacts					
Leads to alcoholism	0.677	0.491	4.866	19.8	0.877
Causes degradation of local heritage	0.747	0.579			
Is likely to increases crime rate	0.700	0.667			
Leads to the alienation between tourists and residents	0.524	0.503			
Is likely to increase drug addiction/trafficking	0.784	0.637			
Causes environmental damage and ecological disruption	0.696	0.522			
Is likely to increase prostitution	0.639	0.507			
Increases vandalism	0.690	0.562			
Negative economic impacts					
Causes more speculative pricing	0.746	0.638	1.252	14.5	0.834
Increases the cost of living	0.799	0.693			
Increases goods and services prices	0.709	0.596			
Causes greater traffic congestion and crowding	0.536	0.551			
Increases property taxes and prices	0.673	0.549			

Source: Authors' elaboration.

Table 12.3: Factor loadings, variance and Cronbach's Alpha of place attachment dimensions of small town destination residents

	Factor loadings	Communalities	Eigen-values	% total variance	Cronbach Alpha
I feel that this place is part of me	0.812	0.660	10.373	64.8	0.963
I prefer this place to any other place	0.843	0.711			
This place is very special to me	0.851	0.724			
I would not trade this place for another	0.778	0.606			
I strongly identify with this place	0.868	0.753			
This place is the best mountain site I know	0.808	0.652			
This place brings back many memories	0.686	0.471			
I don't want to leave this place	0.817	0.667			
I feel I can be myself around this place	0.814	0.662			
Doing what I do here is much more satisfying than doing it elsewhere	0.790	0.624			
I feel very attached to this place and to its people	0.805	0.648			
No other place can be compared to this place	0.770	0.593			
This place says a lot about who I am	0.784	0.615			
Doing what I do in this place is very important to me	0.772	0.596			
This place means a lot to me	0.859	0.738			
I miss this place when I am away	0.809	0.655			

Source: Authors' elaboration.

Correlation

Correlation between tourism impact factors and place attachment was carried out using Pearson Correlation Coefficient (Table 12.4).

The results indicate that there is a positive relationship between residents' place attachment level and their perceptions of the negative and positive social impacts that tourism can have on the place. However, there is no significant relationship between the negative economic impacts and place attachment.

Table 12.4: Correlations between the different impacts of tourism and place attachment according to the residents

		Positive social-cultural impacts	Negative social-cultural impacts	Negative economic impacts	Place attachment
Positive socio-cultural impacts	PearsonCorrelation	1	.000	.000	.302**
	Sig. (2-tailed)		1.000	1.000	.000
	N	300	300	300	300
Negative socio-cultural impacts	Pearson Correlation	.000	1	.000	.134*
	Sig. (2-tailed)	1.000		1.000	.020
	N	300	300	300	300
Negative economic impacts	Pearson Correlation	.000	.000	1	.026
	Sig. (2-tailed)	1.000	1.000		.649
	N	300	300	300	300
Place attachment	Pearson Correlation	.302**	.134*	.026	1
	Sig. (2-tailed)	.000	.020	.649	
	N	300	300	300	300

** Correlation is significant at 0.01 level (2-tailed).

* Correlation is significant at 0.05 level (2-tailed).

Source: Authors elaboration.

Conclusion

According to the results, residents are more sensible to social and cultural impacts, whether positive and negative, and less to negative economic impacts. There is a more positive relationship between residents' place attachment level and their perceptions of the positive socio-cultural impacts compared to the relationship with the negative socio-cultural impacts that tourism may have on the place. However, there is no significant relationship between place attachment and the negative economic impacts. That result could be explained by the strong spirit community that exists in small towns. People are more aware of the social effects that tourism development has on their lives than of its economic or environmental effects. Overall, European small towns have a small and aged population, with fewer job opportunities, poorer health care, and fewer large communication infrastructures. In these communities, people are often eager to meet new people and to connect with other cultures and different generations. Tourists are seen as business opportunities, as people who visit the local bars, restaurants, hotels, and attractions and buy local products to take home with them. On the other hand, tourists are sources of cultural rejuvenation. Some residents have a positive opinion of tourists, whereas for others they represent a negative factor, as they come in and change local customs, habits, and even the local culture.

This study aims to expand existing knowledge on place attachment and residents' perceptions of the impacts of tourism in urban settings. On the other hand, because tourism can be a challenge to small towns and cities, particularly when it comes to their sustainability, the study can aid raise a deeper awareness of the effects that tourism development is having on the community of small town destinations. In addition, the results could help managers by providing long-term economic, social, and cultural benefits to the local community, enhancing quality of life and thus strengthening place and community attachment.

However, this study has some limitations. The study might not consider other specific and relevant existing dimensions of tourism impacts and place attachment. Despite the advantages of the flexibility and easy handling of the structured technique, there is still the risk of omitting important dimensions or using some others that are not so important to the respondents. Moreover, there has been a lack of explanation of why certain impacts are important to or perceived by local residents (Deery, Jago & Fredline, 2012). Qualitative methodology could really help achieve a deeper understanding.

As for future research, it will be interesting to analyze the model's validity in other residential environments, like rural or larger urban destinations. Residents of larger cities or rural places may display different levels of attachment and subsequently different attitudes toward tourism development than smaller towns' residents. On the other hand, and since tourism destinations are dynamic and perceptions change overtime, it will be interesting for researchers to carry out longitudinal studies in order to define a more effective longer-term strategic planning. In addition, since the degree of involvement has been shown to influence residents' overall attitudes toward tourism development (see Ko & Stewart, 2002; Allen, Long, Perdue & Kieselbach, 1988) it could be interesting to analyze whether or not local residents are locally involved in both tourism and non-tourism related activities. Finally, and since the relationship between the perceived qualities of a place and sense of place has been underexplored within the context of tourism development, it could be interesting to relate perception of place, place attachment, and perceived impacts of tourism.

References

Allen, L. R., Long, P. T., Perdue, R. R. & Kieselbach, S. (1988) 'The impact of tourism development on residents', perceptions of community life. *Journal of Travel Research*, 27(1), 16–21.

Almeida-García, F., Peláez-Fernández, M. Á., Balbuena-Vázquez, A. & Cortés-Macias, R. (2016) 'Residents' perceptions of tourism development in Benalmádena (Spain)',*Tourism Management*, 54, 259–274.

Caneday, L. & Zeiger, J. (1991) 'The social, economic, and environmental costs of tourism to a gaming community as perceived by its residents', *Journal of Travel Research*, 30(2), 45–49.

Chen, N. & Dwyer, L. (2018) 'Residents' place satisfaction and place attachment on destination brand-building behaviors: conceptual and empirical differentiation', *Journal of Travel Research*, 57(8), 1026–1041.

Chen, N., Dwyer, L. & Firth, T. (2014a) 'Conceptualization and measurement of dimensionality of place attachment', *Tourism Analysis*, 19(3), 323–338.

Chen, N., Dwyer, L. & Firth, T. (2014b) 'Effect of dimensions of place attachment on residents' word-of-mouth behavior', *Tourism Geographies*, 16(5), 826–843.

Chen, N., Dwyer, L. & Firth, T. (2015) 'Factors influencing Chinese students' behavior in promoting Australia as a destination for Chinese outbound travel', *Journal of Travel & Tourism Marketing*, 32(4), 366–381.

Cronbach, L. J. (1951) 'Coefficient alpha and the internal structure of tests', *Psychometrika*, 16(3), 297–334.

Crouch, G. I. & Ritchie, J. R. B. (1999) 'Tourism, competitiveness, and societal prosperity', *Journal of Business Research*, 44(3), 137–152.

Deery, M., Jago, L. & Fredline, L. (2012) 'Rethinking social impacts of tourism research: a new research agenda', *Tourism Management*, 33(1), 64–73.

Gannon, M., Rasoolimanesh, S. M. & Taheri, B. (2020) 'Assessing the mediating role of residents' perceptions toward tourism development', *Journal of Travel Research*, 1–50.

Garrod, B., Fyall, A., Leask, A. & Reid, E. (2012) 'Engaging residents as stakeholders of the visitor attraction', *Tourism Management*, 33, 1159–1173.

George, R. (2010) 'Visitor perceptions of crime-safety and attitudes towards risk: The case of Table Mountain National Park, Cape Town', *Tourism Management*, 31, 806–815.

Gross, M. J. & Brown, G. (2008) 'An empirical structural model of tourists and places: progressing involvement and place attachment into tourism',*Tourism Management*, 29(6), 1141–1151.

Gu, H. & Ryan, C. (2008) 'Place attachment, identity and community impacts of tourism—the case of a Beijing hutong', *Tourism Management*, 29(4), 637–647.

Gursoy, D. & Rutherford, D. (2004) 'Host attitudes toward tourism: an improved structural model', *Annals of Tourism Research*, 31, 495–516.

Hair, J. F., Black, W., Babin, B. & Anderson, R. (Eds.). (2010). *Multivariate Data Analysis*. Harlow: Pearson.

Hall, C. M. & Page, S. J. (2014). *The Geography of Tourism and Recreation : Environment, Place and Space*. Routledge.

Halpenny, E. A. (2010) 'Pro-environmental behaviors and park visitors: the effect of place attachment', *Journal of Environmental Psychology*, 30(4), 409–421.

Hidalgo, M. C. & Hernández, B. (2001) 'Place attachment: conceptual and empirical questions', *Journal of Environmental Psychology*, 21(3), 273–281.

Hou, J.-S., Lin, C.-H. & Morais, D. B. (2005) 'Antecedents of attachment to a cultural tourism destination: The case of Hakka and Non-Hakka Taiwanese visitors to Pei-Pu, Taiwan'. *Journal of Travel Research*, 44(2), 221–233.

Jaafar, M., Noor, S. M. & Rasoolimanesh, S. M. (2015) 'Perception of young local residents toward sustainable conservation programs: a case study of the Lenggong World Cultural Heritage Site', *Tourism Management*, 48, 154–163.

Jaafar, M., Rasoolimanesh, S. M. & Ismail, S. (2017) 'Perceived sociocultural impacts of tourism and community participation: a case study of Langkawi Island', *Tourism and Hospitality Research*, 17(2), 123–134.

Kim, K., Uysal, M. & Sirgy, M. J. (2013)'How does tourism in a community impact the quality of life of community residents?' *Tourism Management*, 36(C), 527–540.

Knez, I. (2014) 'Place and the self: An autobiographical memory synthesis', *Philosophical Psychology*, 27(2), 164–192.

Ko, D.-W. & Stewart, W. (2002) 'A structural equation model of residents' attitudes for tourism development', *Tourism Management*, 23, 521–530.

Krippendorf, J. (1987) 'Ecological approach to tourism marketing', *Tourism Management*, 8(2), 174–176.

Kyle, G., Bricker, K., Graefe, A. & Wickham, T. (2004) 'An examination of recreationists' relationships with activities and settings', *Leisure Sciences*, 26(2), 123–142.

Kyle, G. T., Absher, J. D. & Graefe, A. R. (2003) 'The moderating role of place attachment on the relationship between attitudes toward fees and spending preferences', *Leisure Sciences*, 25(1), 33–50.

Látková, P. & Vogt, C. A. (2012) 'Residents' attitudes toward existing and future tourism development in rural communities', *Journal of Travel Research*, 51(1), 50–67.

Lawton, L. J. & Weaver, D. B. (2015) 'Using residents' perceptions research to inform planning and management for sustainable tourism: a study of the Gold Coast Schoolies week, a contentious tourism event', *Journal of Sustainable Tourism*, 23(5), 660–682.

Lewicka, M. (2011) 'Place Attachment: how far have we come in the last 40 years?', *Journal of Environmental Psychology*, 31, 207–230.

Liu, J., Sheldon, P., & Var, T. (1987) 'Cross-national approach to determining resident perceptions of the impact of tourism on the environment', *Annals of Tourism Research*, 14 (1), 17-37.

Low, S. M. & Altman, I. (1992) 'Place attachment', in: Altman, I. and Low, S. M. (Eds.), *Place Attachment*, (pp. 1–12). Boston, MA: Springer US. Retrieved January 23, 2020, from https://doi.org/10.1007/978-1-4684-8753-4_1.

Lundberg, E. (2017) 'The importance of tourism impacts for different local resident groups: A case study of a Swedish seaside destination', *Journal of Destination Marketing & Management*, 6(1), 46–55.

Manzo, L. C. (2003) 'Beyond house and haven: toward a revisioning of emotional relationships with places', *Journal of Environmental Psychology*, 23(1), 47–61.

Mason, P. (2003). *Tourism Impacts, Planning and Management.* Oxford: Butterworth-Heinemann.

Mathieson, A. & Wall, G. (1982) *Tourism: Economic, Physical, and Social Impacts*. London: Longman.

Monterrubio, C. (2016) 'The impact of spring break behaviour: an integrated threat theory analysis of residents' prejudice', *Tourism Management*, 54, 418–427.

Moore, R. L. & Scott, D. (2003). 'Place attachment and context: comparing a park and a trail within', *Forest Science*, 49, 877–884.

Nepal, S. K. & Chipeniuk, R. (2005) 'Mountain tourism: toward a conceptual framework', *Tourism Geographies*, 7(3), 313–333.

Nunkoo, R. & Ramkissoon, H. (2010) 'Small island urban tourism: a residents' perspective', *Current Issues in Tourism*, 13(1), 37–60.

Prayag, G. & Ryan, C. (2012)'Antecedents of tourists' loyalty to Mauritius: The role and influence of destination image, place attachment, personal involvement, and satisfaction', *Journal of Travel Research*, 51(3), 342–356.

Proshansky, H. M., Fabian, A. K. & Kaminoff, R. (1983) 'Place-identity: Physical world socialization of the self', *Journal of Environmental Psychology*, 3(1), 57–83.

Ramkissoon, H. & Mavondo, F. T. (2015) 'The satisfaction–place attachment relationship: potential mediators and moderators', *Journal of Business Research*, 68(12), 2593–2602.

Ramkissoon, H., Weiler, B. & Smith, L. D. G. (2012) 'Place Attachment and pro-environmental behaviour in national parks: the development of a conceptual framework', *Journal of Sustainable Tourism*, 20(2), 257–276.

Rasoolimanesh, S. M., Ali, F. & Jaafar, M. (2018) 'Modeling residents' perceptions of tourism development: linear versus non-linear models', *Journal of Destination Marketing & Management*, 10, 1–9.

Rasoolimanesh, S. M., Badarulzaman, N., Abdulah, A. & Behrang, M. (2019) 'How governance influences the components of sustainable urban development',*Journal of Cleaner Production*.

Rasoolimanesh, S. M., Jaafar, M., Kock, N. & Ramayah, T. (2015) 'A revised framework of social exchange theory to investigate the factors influencing residents' perceptions', *Tourism Management Perspectives*, 16, 335–345.

Rasoolimanesh, S. M., Ringle, C. M., Jaafar, M. & Ramayah, T. (2017) 'Urban vs. rural destinations: residents' perceptions, community participation and support for tourism development', *Tourism Management*, 60, 147–158.

Ratcliffe, E. & Korpela, K. M. (2016) 'Memory and place attachment as predictors of imagined restorative perceptions of favourite places', *Journal of Environmental Psychology*, 48, 120–130.

Raymond, C. M., Brown, G. & Weber, D. (2010) 'The measurement of place attachment: personal, community, and environmental connections', *Journal of Environmental Psychology*, 30(4), 422–434.

Šegota, T., Mihalič, T. & Kuščer, K. (2017)'The impact of residents' informedness and involvement on their perceptions of tourism impacts: the case of Bled', *Journal of Destination Marketing & Management*, 6(3), 196–206.

Sharma, S. (1996). *Applied Multivariate Techniques*. New York: J. Wiley.

Sharpley, R. (2014) 'Host Perceptions of Tourism: A Review of the Research', *Tourism Management*, 42, 37–49.

Shumaker, S. & Taylor, R., B. (1983) 'Toward a clarification of people-place relationship: a model of attachment to place', in N.R. Feimer (ed.), *Environmental Psychology. Directions and Perspectives* (pp. 12571). London: Sage Publications. Retrieved 23/01/2020, from http://www.europhd.net/bibliographic-item/toward-clarification-people-place-relationship-model-attachment-place.

Starr, N. (2002). *Viewpoint: An Introduction to Travel, Tourism, and Hospitality*. NJ: Prentice Hall.

Stylidis, D. (2018) 'Place attachment, perception of place and residents' support for tourism development', *Tourism Planning & Development*, 15(2), 188–210.

Thompson-Fawcett, M. & McGregor, S. (2011) 'Tourism in a small town: impacts on community solidarity', *International Journal of Sustainable Society*, 3, 174–189.

Tsai, S. (2012) 'Place attachment and tourism marketing: investigating international tourists in Singapore', *International Journal of Tourism Research*, 14(2), 139–152.

Vorkinn, M. & Riese, H. (2001). Environmental concern in a local context: the significance of place attachment', *Environment and Behavior*, 33(2), 249–263.

Wang, Y. (Alex) & Pfister, R. E. (2008) 'Residents' attitudes toward tourism and perceived personal benefits in a rural community'. *Journal of Travel Research*, 47(1), 84–93.

West, M. . (1991). *Modelling with Mixtures*. Duke University.

Williams, D. R., Patterson, M. E., Roggenbuck, J. W. & Watson, A. E. (1992)'Beyond the commodity metaphor: examining emotional and symbolic attachment to place', *Leisure Sciences*, 14(1), 29–46.

Woosnam, K. M., Aleshinloye, K. D., Ribeiro, M. A., Stylidis, D., Jiang, J. & Erul, E. (2018) 'Social determinants of place attachment at a world heritage site', *Tourism Management*, 67, 139–146.

Yuksel, A., Yuksel, F. & Bilim, Y. (2010) 'Destination attachment: effects on customer satisfaction and cognitive, affective and conative loyalty', *Tourism Management*, 31(2), 274–284.

13 Conclusion: Preparing for the future of travel and tourism in vulnerable times

Alain Decrop and Antónia Correia

When gathering the best papers presented in 2019 during the 8th ATMC conference in Namur for editing this book, we were not expecting 2020 to become the nightmare year that we all experienced in the world. The Covid-19 crisis has significantly disrupted our consumption and travel habits. Heaps of business sectors were severely impacted by the two waves of lockdown of populations and countries, the closure of stores and restaurants and the overall context of anxiety and uncertainty. More than other sectors, the tourism and travel industry was severely affected. The trips and recreative activities of our fellow citizens are henceforth punctuated by 'barrier gestures' and by a large number of health measures taken in order to limit the spread of the pandemic.

Of course, the objective of this book was not to account for the current crisis, nor to analyze the extent to which our tourism and leisure activities have changed, and whether these changes are likely to last. The aim of the book was first to discuss sustainability and collaborative practices in order to fill tourism development with ethics and responsibility. Following the collegial tradition of Advances in Tourism Marketing conferences, this book is the result of the collaboration of established international scholars with emergent researchers around the globe. The twelve chapters of this book have offered innovative and dovetailing perspectives around four major strategic questions that may support tourism development in these vulnerable times, i.e.: Can technology enhance value co-creation and the tourism experience? How sharing economy practices develop in tourism? To what extent sustainability issues affect culture and the well-being of residents? What is the connection between technology, millennials and overtourism?

Of course, those issues were considered in the framework of the digital world that is ever expanding since two decades. Digital technology has been massively integrated into the strategies and operations of tourism operators. Historically dedicated to marketing and communication, digital technology is now revolutionizing the entire tourism value chain. The promotion of a destination can no longer be considered without the mix between official digital channels and levers such as influencers and rating platforms. Any tourist expects to find relevant and up-to-date information on the Internet in order to prepare for his stay, to choose among transportation, accommodation and recreational alternatives, and to book activities. The exchange of tips and advice has become an important element in choosing a destination. In many regions, competition between destinations and operators makes it essential to be attractive via a different offer and digital technology may act as a major differentiator (e.g., Histopad to visit Chambord's Castle, Monument Tracker).

To answer these questions, the book was organized in four parts where interdependence, collaboration, sustainability and responsible behavior are deeply discussed with the aim of contributing to marketing tourism in the digital world that we know today. The book started with a discussion about how to co-create value with or without technologies, detailed how web platforms reboot collaborative economic practices, presented sustainable tourism development under a human perspective and ended with the discussion of key issues in tourism marketing, including overtourism and residents' propensity to boost tourism development. We now summarize the key takeaways of those four parts, closing each part with a reflection on the lessons learnt from current Covid-19 crisis in order to assess what will be left after the vulnerable times that we live now.

1. Acceleration of digital transformation

We have observed a technological and digital transformation both on consumers' and on providers' sides for several years. Online booking and electronic means of payment are getting more and more popular. The development and adoption of smart technologies, including artificial intelligence (AI), robotics, and automation, in the travel and tourism sector is accelerating.

The first part of the book was devoted to the role of such technologies in enhancing value co-creation and tourist experiences. Chapter 2 explored the research streams on absorptive capacity, co-creation and tourism. The state of the art performed recalls for the alignment of these three cornerstones that help to connect a number of research trends: the reciprocity tourism industry needs to acquire, the employment that needs to be valorized, the use of social media to enhance digital transformation in the sector, and the stakeholders' sense-making and sense-giving process. This chapter, with a macro perspective, offers paths for future research challenges, such as opening tourism to the absorptive black box in co-creation experiences within a framework that involves all stakeholders, including universities. This chapter suggests that interdependence is key in order to co-create knowledge. In line with this, Chapter 3 explored a non-western perspective on how to co-create tourism experiences through social interactions. Surprisingly, Chinese tourists aim to interact with service providers and with locals to enhance quality and authenticity to their experience but they tend to avoid interacting with other tourists, which could be an advantage in Covid times. More than any other tourism and leisure sectors, the festival and entertainment industry was severely damaged by the crisis as most events have been cancelled and activities have been stopped, because the gathering of crowds is forbidden. Chapter 4 showed that social and emotional interactions lived during and after the festival experience are determinants of consumers' satisfaction. Festivals involve a collective and collaborative experience where emotional contagion occurs at interpersonal and mass levels.

The lockdown linked to the Covid-19 pandemic has accelerated the adoption of digital technologies by consumers and amplified digital innovation efforts on the part of travel operators and tourist sites. Many initiatives have been carried out to develop online booking and visit planning in order to allow better management of flows, or to highlight health compliance operations via their websites and social networks in order to reassure and prepare the return of visitors. Many tourist sites have set up interactive platforms and virtual tours in order to keep in touch with visitors. The theatricalization of confinement has made it possible to communicate on social networks via strong images presenting nature and authenticity regaining possession of empty places. The digital has made it possible to maintain emotional connections between festivals and their consumers, as illustrated by the case of Tomorrowland and other artists who have set up online music events.

2. Mutations in the sharing economy

Another major trend of these last years is the development of the so called 'sharing' or 'collaborative' economy, which is another illustration of the importance of interdependence and technology particularly affecting tourism. This second part of the book started with the discussion of the legal issues that stem from collaborative practices, before turning to the presentation of examples of how different platforms provide tourism experiences in that collaborative economy. Chapter 5 showed that most of the sharing platforms are not aligned with the consumer law of the European Union. It seems that Airbnb is one of the few exceptions obeying consumer rights' regulations. The other platforms have space to improve in what concern transparency. How tourists use these platforms was discussed in the following chapters. Chapter 6 presented sharing meals' experiences through the EatWith platform. This study identifies serendipity as a moderator of meal-sharing experiences and conclude that serendipity in meal-sharing experiences included the hospitality, the food and the tourist-tourist interaction. Chapter 7 compared the best known collaborative platforms, AirBnB and Couchsurfing, that diverge in terms of business models. In fact, consociality seems to be more prominent in Couchsurfing than in Airbnb, suggesting that monetized exchanges hinder the true sense of sharing and disturb collaborative initiatives where social interactions are the core of the business.

The current Covid-19 crisis has impacts on shareable tourism. Like many other businesses, platforms such as Airbnb, Uber or Blablacar have suffered from the lockdown of countries and populations whereas other players such as Deliveroo or Uber Eats have benefited from the confinement situation that led more consumers to have food delivered to their homes. However it is likely that collaborative initiatives in tourism and travel will keep on growing in the post-pandemic years, as peer-to-peer exchanges empower consumers and providers of services, help them to get some economic benefits in difficult times and to restore trust and hope for a better future. As matchmakers moderating encounters between hosts and guests, sharing platforms offer more direct, consocial and authentic experiences than the conventional industry. These platforms may help tourism to rebound in more responsible and sustainable patterns of development.

3. Return to the local and to nature

Sustainability was explored in Part III of this book through the presentation of two chapters that discussed sustainable tourism development. Chapter 8 extended the theory of social exchange to the field of sustainability; host-tourists' interactions are considered to influence residents' attitude and the supportiveness of residents towards sustainable tourism development depends on tourist attitudes. Chapter 9 explored how government and stakeholders should be coordinated to ensure sustainable tourism development in heritage places. The chapter concluded that management and planning are mandatory to develop tourism on sustainable grounds, nevertheless the blueprint is cooperation. These two chapters showed that sustainability is likely to impact residents' wealth and well-being and that cooperation among stakeholders is key for a sustainable development.

Of course, eco-tourism and sustainability issues are not new in tourism but the current crisis is likely to reinforce the return to the local and the nature. Covid-19 is a disaster that has shown the limitations of global economic models and has strongly restricted international travel possibilities. In the last few months, 'locavorism' has been the new trend in town, characterized by the evolution towards short circuits, reduced travel and transport practices, organic local food, regional or domestic tourism, staycation and micro-vacations. In parallel, there is a craze for 'green' tourism and eco-tourism: slow tourism far from crowds and cities, visits to natural sites, stays in (glamorous) campings and rural accommodation, roadtrips in complete independence to avoid any risk of contamination etc.

4. The undermining of mass-tourism models

The last part of the book presented three cases where tourism development is compromised. Chapter 10 explored how digital communication may increase overtourism and concluded that controlling tourist-generated content can help to reposition a tourism destination in more sustainable patterns. In line with this, Chapter 11 analysed residents' perceptions about overtourism induced by cruises in cities such as Venice. For residents, the negative effects of mass cruises significantly outweigh the positive ones. This result suggests that the evidence of negative externalities shape residents to be against tourism growth, which recalls once more for the need of sustainability. Chapter 12 explored place attachment and residents' perceptions of the impacts of tourism in urban settings. According to this study, residents are more sensitive to social and cultural impacts, whether positive and negative, and less to negative economic impacts.

Residents seem to be more aware of the social effects that tourism development has on their lives than of its economic or environmental effects. The authors explained this evidence by loneliness and the cultural rejuvenation brought by tourism.

During the last months, overtourism that was a 'hot topic' in the last years has completely disappeared as a consequence of the travel barriers and restrictions dictated by governments. The question is now: what will happen next...? We may expect that changes in the decision-making processes such as risk aversion, increased planning of activities, trips and holidays, online booking etc. are likely to last for a long time. The Covid-19 pandemic has undermined a number of business models related to long-haul flights, cruises, organized trips, low-cost airlines, business tourism, cultural events etc. Such industries are likely to suffer more from the consequences of the crisis as they depend heavily on large numbers of customers being gathered in curtailed spaces to be profitable. However, one should not forget that overtourism is mostly induced by communication and promotion, especially through the viral effect of social networks. The vulnerable times we know today should be used to redraft tourism fluxes by a replanning of communication efforts and user-generated contents. Furthermore, a number of destinations take benefit of the situation to rethink their tourism strategies, in order to make them more sustainable and/or to show the positive externalities of tourism to their residents, in particular as to cultural and social interactions.

The months of planetary containment had a cataclysmic effect on the tourism sector. While it is difficult to know what will happen at the end of the looming economic crisis, one can imagine that, after a long and painful period of transition, our developed societies will enter a new world – a world we hope to be better and to tackle climate change. Some believe that the crisis will radically transform mobility on a global scale, that mass tourism has fizzled out and that the world after will reinvent the relationship to travel, places and people, emphasizing the happiness of proximity and slowness. At the same time, alternatives to mass tourism, based on new values (local roots, associative commitment, personal development, etc.) should be developed. Implicit strategies and the cases presented along this book show that in these vulnerable times, we should learn how to (re)live together, even with more distant social interactions, if we do not want to perish together.

The Covid-19 crisis has amplified a number of paradoxical tensions that we already observed in the last years and that have paved the way for conferences such as ATMC and an edited book like this one. The society we live in is more and more digitalized and virtualized, but at the same time characterized by a growing desire of many to restore straight and genuine human connections. An ever increasing number of consumers are shopping globally on the Internet but meanwhile, we also notice the return to local consumption. A last paradox is that, on the one hand, sustainability and caring for the planet have become central values for many tourists but, on the other hand, most of them have never been so sensitive to price and saving money. On top of such paradoxical tensions, maybe the major challenge for the tourism industry will be to restore trust and (physical and psychological) well-being after this long period of unparalleled fear and anxiety.

Index

www.ingramcontent.com/pod-product-compliance
Ingram Content Group UK Ltd.
Pitfield, Milton Keynes, MK11 3LW, UK
UKHW021828100926
8556IPUK00003B/225

9 781911 635765